ELEMENTARY
PHYSICAL CHEMISTRY

ELEMENTARY
PHYSICAL CHEMISTRY

Bruno Linder

Florida State University, USA

World Scientific

NEW JERSEY • LONDON • SINGAPORE • BEIJING • SHANGHAI • HONG KONG • TAIPEI • CHENNAI

Published by

World Scientific Publishing Co. Pte. Ltd.

5 Toh Tuck Link, Singapore 596224

USA office: 27 Warren Street, Suite 401-402, Hackensack, NJ 07601

UK office: 57 Shelton Street, Covent Garden, London WC2H 9HE

British Library Cataloguing-in-Publication Data
A catalogue record for this book is available from the British Library.

ELEMENTARY PHYSICAL CHEMISTRY

ISBN-13 978-981-4299-66-4

Typeset by Stallion Press
Email: enquiries@stallionpress.com

Printed in Singapore.

To Cecelia
... and to William, Diane, Richard, Nancy, and Carolyn

Preface

This book is based on a set of lecture notes for a one-semester course in general physical chemistry (CHM 3400 at Florida State University) taught by me in the Spring of 2009. This course was designed specially for students working towards a Baccalaureate degree in Chemical Science. The course, entitled *General Physical Chemistry*, consisted of three lectures weekly and one recitation hour meeting weekly.

The course consisted of an elementary exposition of general physical chemistry and included topics in thermodynamics, phase and chemical equilibria, cell potentials, chemical kinetics, introductory quantum mechanics, elements of atomic and molecular structure, elements of spectroscopy, and intermolecular forces.

In the field of physical chemistry, especially thermodynamics and quantum mechanics, there are subtleties and conceptual difficulties, often ignored in even more advanced treatments, which tend to obscure the logical consistency of the subject. While the emphasis in this course is not on mathematical rigor, conceptual difficulties are not "swept under the rug", but brought to the fore.

An essential feature of the course is weekly assignment of homework problems, reflecting more or less the topic contents. These problems were graded and discussed by the assigned teaching assistant in the recitation session.

It is a pleasure to thank Jared Kinyon for reading the proofs and checking the problems, and Steve Leukanech for doing the drawings.

Bruno Linder
August 2010

Contents

Chapter 1

State of Matter. Properties of Gases

Chemistry deals with the *properties* of matter, the *changes* matter undergoes and the *energy* that accompanies the changes.

Physical Chemistry is concerned with the *principles* that underlie chemical behavior, the *structure* of matter, *forms* of energy and their interrelations and *interpretation of macroscopic* (bulk) properties of matter in terms of their *microscopic* (molecular) constituents.

Broad classification of Matter: A **gas** fills the container and takes on the shape of the container. A **liquid** has a well-defined surface and a fixed volume but no definite shape. A **solid** has a definite shape, a fixed volume, and is independent of constraints.

The foregoing classification is a macroscopic classification. From a microscopic (molecular) point of view — a **gas** consists of particles that interact with each other weakly; a **liquid** consists of particles that are in contact with each other but are able to move past each other; and a **solid** consists of particles that are in contact with each other but are unable to move past each other. For short, in a gas, particles have essentially no restriction on motion; in a solid, particles are locked together, mostly with fixed orientation; and in liquid, particles behave in a manner between gas and solid.

1.1. State of Matter

The above classification is often referred to as a classification into states *of aggregation*. In physical chemistry, the word *state* generally refers to another concept. A substance is described by its properties (pressure, volume, temperature, amount, composition, etc.). If all the properties

1

of a substance are specified, the **state** of the system is said to be specified. Actually, there is no need to specify *all* properties, because, as a rule, the properties are interdependent. For example, if you know the pressure, the volume and the number of moles n of an ideal gas, you can figure out the temperature from the equation of state: $[PV = nRT]$.

1.2. Description of Some States of Matter

- Volume, V: a measure of occupied space.
- Pressure: force per unit area.
- Temperature: hard to define rigorously,* but in simple language it is a measure of the degree of hotness or coolness for which all of us have an intuitive feeling.
- Amount of substance: a measure of the amount of matter.

*Comment: When two objects (bodies) are brought in contact with each other, the hotter body will cool, the colder body will heat up. This is interpreted that *heat (a form of energy) is flowing from the hotter body to the colder one.* This process will continue until no more heat is transferred. When that happens the two bodies are said to be in *thermal equilibrium* — and the temperatures of the two bodies will be the same.

1.3. Units

The recommended units are SI (Systeme Internationale) units:

Length	l	meter, m
Mass	m	kilogram, kg
Time	t	second, s
Electric current	I	ampere, A
Temperature	T	Kelvin, K
Amount	n	mole, mol

All other physical quantities that we use can be derived from these. For example, volume is length cube or m^3. Some derived quantities have special names. For example,

- Force in SI units is kg m s^{-2} or Newton, N.
- Pressure in SI units is kg m^{-1}s^{-2} or pascal, Pa.
- Energy in SI units is kg m^2 s^{-2} or joule, J.

Other (non-SI) units frequently used are:

- Pressure: mmHg or Torr (1 Torr = 133.3 Pa) or atm (1 atm = 760 mmHg) or 101.325 kPa bar (1 bar = 10^5 Pa)
- Energy: electron volt, eV (1 eV = 1.602×10^{-19} J)

Equation of state is an algebraic relation between pressure, volume, temperature and quantity of substance.

1.4. Ideal or Perfect Gas Law

$$PV = nRT$$

This Law comprises three different Laws that preceded it.

1) Holding constant n and T gives PV = const or

$$P_1 V_1 = P_2 V_2 \qquad \text{Boyle's Law} \qquad (1.1)$$

2) Holding constant P and n gives V/T = const or

$$V_1/T_1 = V_2/T_2 \qquad \text{Charles' Law} \qquad (1.2)$$

3) Holding constant P and T gives V/n = const or

$$V_1/n_1 = V_2/n_2 \qquad \text{Avogadro's Law} \qquad (1.3)$$

1.5. Evaluation of the Gas Constant, R

The gas constant can be expressed in various units, all having the dimension of energy per degree per mol.

a) R is most easily calculated from the fact that the hypothetical volume of an ideal gas is 22.414 L at STP (273.1 K and 1 atm). Accordingly,

$$R = (1\,\text{atm})(22.414\,\text{L mol}^{-1})/(273.16\,K)$$
$$= 0.08206\,\text{atm L K}^{-1}\text{mol}^{-1} \qquad (1.4)$$

b) If V is in cm^3,

$$R = (1\,\text{atm})(22,414\,\text{cm}^3\,\text{mol}^{-1})/(273.16\,K)$$
$$= 82.06\,\text{atm cm}^3\text{K}^{-1}\,\text{mol}^{-1} \qquad (1.5)$$

c) In Pascal L K^{-1} mol^{-1} [1 atm = 1.01325×10^5 Pa; 1 L= 10^{-3} m^3],

$$R = 1.01325 \times 10^5 \, \text{Pa} \times 22.414 \times 10^{-3} \, \text{m}^3 \, \text{mol}^{-1}/273.16 \, \text{K}$$
$$= 8.314 \, \text{Pa} \, \text{m}^3 \, \text{K}^{-1} \, \text{mol}^{-1}$$
$$= 8.314 \, \text{k Pa L K}^{-1} \, \text{mol}^{-1} \tag{1.6}$$

d) In J K^{-1} mol^{-1},

$$R = 8.314 \, \text{kg} \, \text{m}^2 \text{s}^{-2} = 8.314 \, \text{J K}^{-1} \, \text{mol}^{-1} \, [1 \, \text{Pa} = 1 \, \text{kg} \, \text{m}^{-1} \, \text{s}^{-2}] \tag{1.7}$$

e) In cgs units (V in cm^3, P in dyne/cm^2, 1 atm = 1.013×10^6 dyne cm^{-2}),

$$R = (1.013 \times 10^6 \, \text{dyne cm}^{-2}) \times (22,414 \, \text{cm}^3 \, \text{mol}^{-1})/273.16 \, \text{K}$$

Also 1 erg = 10^7 J,

$$R = 8.314 \times 10^7 \, \text{erg K}^{-1} \, \text{mol}^{-1} \tag{1.8}$$

f) In cal K^{-1} mol^{-1} (1 cal = 4.184 J),

$$R = 1.987 \, \text{cal K}^{-1} \text{mol}^{-1} \tag{1.9}$$

Example 1.1. 50.0 g of N$_2$ (M = 28.0 g) occupies a volume of 750 mL at 298.15 K. Assuming the gas behaves ideally, calculate the pressure of the gas in kPa.

Solution

$P = nRT/V$

$\quad = (50.0 \, \text{g}/28.0 \, \text{g} \, \text{mol}^{-1}) \times (0.0826 \, \text{atm L K}^{-1} \text{mol}^{-1} \times 298.15 \, \text{K})/0.750 \, \text{L}$

$\quad = 58.25 \, \text{atm} = 58.25 \times 101.325 \, \text{kPa/atm} = 5.90 \times 10^3 \, \text{kP}$

1.6. Mixtures of Gases

The partial pressure of a gas in a mixture is defined as the pressure the gas would exert if it *alone* occupied the whole volume of the mixture at the same temperature. *Dalton's Law* states that the total pressure is equal to the sum of the partial pressures. That is,

$$P = \Sigma_i P_i = \Sigma_i n_i RT/V = RT\Sigma_i n_i/V = nRT/V \qquad (1.10a)$$

where $n = \Sigma_i n_i$ is the total number of moles. Accordingly,

$$P_i/P = n_i/n = x_i \quad \text{or} \quad P_i = x_i P \qquad (1.10b)$$

[This relation is strictly valid for ideal gases.]

1.7. The Kinetic Theory of Gases

The theory is based on the following assumptions:

1. There are N molecules, each of weight m.
2. Molecules are in constant motion. They collide with each other and with the walls of the container.
3. In ideal gases, molecules do not interact with each other.
4. The volume of molecules is negligible compared to the volume of container.

Consider one molecule in a cubic box colliding with a shaded wall parallel to the YZ direction. Before collision, the velocity of molecule in the X-direction is u_x. When the molecule collides with the shaded wall (see Fig. 1.1) of the cubic box, it is reflected in the opposite direction, having a velocity of $-u_x$ and a change of velocity of $2u_x$. If the distance between the shaded wall and

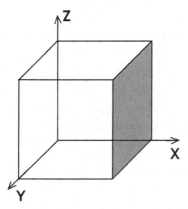

Fig. 1.1 Depicts a particle in a cubic box of sides L colliding with the shaded wall.

the opposite wall is L, the molecule is reflected in the opposite direction, having a velocity of $-u_x$. and a change of velocity of $2u_x$. The molecule will make $u_x/2L$ collisions per unit time with the shaded wall. Accordingly, the change in momentum per molecule per unit time at the shaded wall will be $(2mu_x) \times (u_x/2L) = mu_x^2/L$. For N molecules, the change in momentum per unit time will be $Nm\langle u_x^2 \rangle/L$ where $\langle \rangle$ stands for average.

In classical mechanics, the momentum change on an area represents the force exerted on that area. Denoting the force as f we can write $f = Nm\langle u_x^2 \rangle/L$ as the force exerted on the shaded wall. Pressure is force per unit area, $P = f/A$, and so $P = Nm\langle u_x^2 \rangle/V$, where V is the volume $V = A \times L$. This oversimplified analysis shows how a macroscopic (thermodynamic) property, i.e. pressure, can be related to the microscopic (mechanical) property, i.e. molecular velocity.

Thus,

$$P = f/A^2 = Nm\langle u_x^2 \rangle/L^3 \tag{1.11}$$

If c denotes the speed in 3-dimensions, $c^2 = u_x^2 + u_y^2 + u_z^2$, we can write $\langle u_x^2 \rangle = \frac{1}{3}\langle c^2 \rangle$, yielding

$$P = \frac{1}{3}Nm\langle c^2 \rangle/L^3 = \frac{1}{3}Nm\langle c^2 \rangle/V \tag{1.12a}$$

If N_A is Avogadro's number, then $Nm = nN_A m = nM$, where M is the molar mass. Thus,

$$PV = \frac{1}{3}nM\langle c^2 \rangle \tag{1.12b}$$

Equating this to the ideal gas law gives, for $n = 1$, the root-mean-square velocity:

$$c_{rms} = c = \sqrt{\langle c^2 \rangle} = \sqrt{(3RT/M)} \tag{1.12c}$$

Conclusion: *The root-mean square speed of a molecule in an ideal gas is proportional to the square-root of the temperature and inversely proportional to its mass.*

Example 1.2. What is the mean square speed of a N_2 molecule (treated as an ideal gas) at a temperature of 25°C and a pressure of 1 bar (10^5 Pa)?

Using $R = 8.314 \, \text{Pa} \, \text{m}^3 \, \text{K}^{-1} \, \text{mol}^{-1}$ and observing that $1 \, \text{Pa} = 1 \, \text{kg} \, \text{m}^{-1} \text{s}^{-2}$ and that the molar mass of N_2 is $M = 28.0 \, \text{g/mol}$ or

28.0×10^{-3} kg/mol gives

$$c_{rms} = (3 \times 8.3145 \times \mathrm{kg\, m^2 s^{-2}\, K^{-1} mol^{-1}} \times 298\,\mathrm{K}/28 \times 10^{-3}\,\mathrm{kg\, mol^{-1}})^{1/2}$$

$$= 515.2\,\mathrm{ms^{-1}}$$

We now have a relation between the macroscopic quantity T and the microscopic property, c. Since the average energy of molecule is $\langle \varepsilon \rangle = \frac{1}{2} m \langle c^2 \rangle$ we immediately obain

$$PV = \frac{2}{3} N \langle \varepsilon \rangle \qquad (1.13a)$$

and for one mole,

$$PV = RT = \frac{2}{3} N_A \langle \varepsilon \rangle \qquad (1.13b)$$

where N_A is Avogadro's number. Finally,

$$\langle \varepsilon \rangle = \frac{3}{2} RT/N_A \qquad (1.13c)$$

Defining R/N_A as Boltzmann's constant ($k = 1.38 \times 10^{-16}\,\mathrm{erg\, K^{-1}}$ molcule^{-1}) gives

$$\langle \varepsilon \rangle = \frac{3}{2} kT \qquad (1.14)$$

Comment 1: Temperature is not associated with the kinetic energy of a single molecule, but with the average kinetic energy of a large number of molecules. It is a statistical concept.

Comment 2: So far we have dealt only with average speeds. Actually, the speeds of molecules vary enormously. Molecules slow down as they collide with one another, speed up afterwards, etc. An expression of the distribution of speeds was derived by *Maxwell*. A schematic diagram of the variation of speed with temperature is depicted in Fig. 1.2.

1.8. Molecular Collisions

The *mean free path*, λ, is the average distance (of molecules) between collisions. The *collision frequency*, z, is the rate at which single molecule

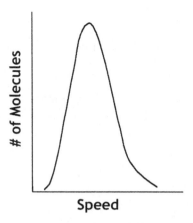

Fig. 1.2 Variation of the number of molecules with speed.

collides with other molecules (i.e. number of collisions/second). It is obvious
that the root-mean-square speed is $c = \lambda z$.

The kinetic theory developed so far cannot be used to derive λ. We must
take into account the finite size of the particles. The result (not derived
here) is

$$\lambda = RT/(\sqrt{2}N_A\sigma P) \tag{1.15}$$

$$z = \sqrt{2}N_A\sigma cP/(RT) \tag{1.16}$$

where σ is the area, $\sigma = \pi d^2$, and d the diameter.

Note:

1) $\lambda z = c$.
2) When P increases, λ increases and z increases.
3) Gases with larger σ have smaller λ and greater z.

1.9. Diffusion of Gases. Graham's Law

Diffusion is the tendency of a substance to spread uniformly through space
available to it. Effusion is escape of a gas through a small hole.

The rate at which gases diffuse depends on the density. Graham's Law
states that the rate of diffusion is inversely proportional to the square root
of the density. If D_1 and D_2 represent the rate of diffusion of Gas 1 and
of Gas 2, Graham's Law suggests that the rate of diffusion is inversely

proportional to the square root of the density, ρ. That is

$$D_1/D_2 = \sqrt{(\rho_2/\rho_1)} \tag{1.17}$$

Since for an ideal gas $PV = (m/M)RT$, where m is the total mass of the gas and M its molecular weight, we can write

$$P = (m/V)(RT/M) = (\rho/M)RT \tag{1.18}$$

It follows that for two gases at a given P and T,

$$\rho_2/\rho_1 = M_2/M_1 \tag{1.19}$$

and thus

$$D_1/D_2 = \sqrt{(M_2/M_1)} \tag{1.20}$$

1.10. Molecular Basis of Graham's Law

It is natural to suppose that the rate of diffusion is proportional to the root-mean-square velocity, that is, D is proportional to $\sqrt{\langle c^2 \rangle}$ or to c. Accordingly,

$$D_1/D_2 = c_1/c_2$$
$$= \{\sqrt{(3RT/M_1)}/\sqrt{(3RT/M_2)}\}$$
$$= \sqrt{M_2/M_1} \tag{1.21a}$$

It follows also that for the same gas at different temperatures,

$$D_1/D_2 = \sqrt{(T_1/T_2)} \tag{1.21b}$$

and for different gases at the same P and T,

$$D_1/D_2 = \sqrt{(M_2/M_1)} \tag{1.21c}$$

in accordance with the kinetic theory of gases.

Example 1.3.

a) Calculate the root-mean-square speed (in ms^{-1}) of a H_2 molecule at $T = 298.15\,K$. The root-mean-square speed is $c = \sqrt{(3RT/M)}$. Taking R in Joule ($1\,J = kg\,m^2 s^{-2}$) and M in $kg\,mol^{-1}$, we get

$$c = \sqrt{(3 \times 8.3145\,kg\,m^2 s^{-2} \times 298.15\,K/0.028\,kg\,mol^{-1})} = 515.4\,m\,s^{-1}$$

b) Calculate the ratio of the rate of diffusion of

$$O_2(M = 32.0\,\mathrm{g\,mol}^{-1}) \text{ to } N_2(M = 28.0\,\mathrm{g\,mol}^{-1})$$

The ratio is proportional to the ratio of the speeds of the molecules, and if P and T are the same for both molecules,

$$(D \text{ of } O_2/D \text{ of } N_2) = c\,(O_2)/c(N_2) = \sqrt{[3RT/M(O_2)]/[3RT/M(N_2)]}$$
$$= \sqrt{[M(N_2)/M(O_2)]} = \sqrt{(28.0/32.0)} = 0.95$$

1.11. Real Gases

So far attention was focused on ideal gases. From a molecular point of view, ideal gases consist of molecules that do not attract or repel each other. This is obviously unrealistic. In a real gas (even if the molecules have no dipoles, quadrupoles, etc. or electrical charges), there are short-range repulsive forces and long-range attractive forces, which invalidates the ideal equation of state.

An equation of state that takes into account these interactions is the

a) **van der Waals equation of state**

$$(P + an^2/V^2)(V - nb) = n\,RT \tag{1.22}$$

where a and b are constants.

Another equation of state is the

b) **Virial equation of state**

$$PV_{\mathrm{m}} = RT[1 + B/V_{\mathrm{m}} + C/V_{\mathrm{m}}^2 + \cdots] \tag{1.23}$$

where V_{m} is the molar volume of the gas, B the *second* virial coefficient, C the *third* virial coefficient, etc.

Attractive forces are needed to account for liquefaction of gases. When a compressed gas in a container is forced through a porous plug into another where it is less compressed (the Joule–Thomson Experiment), the gas cools. Why? In the compressed state the molecules are close to each other; there is great attraction. In the dilute state, the molecules are farther apart. Therefore, when the gas expands the attractive *van der Waals bonds* are broken. It takes energy to do that. The energy comes from the gas — the gas cools!

Chapter 2

The First Law of Thermodynamics

Thermodynamics is concerned with the laws that govern the transformation of one kind of energy into another during a physical or chemical change.

The term **energy** is difficult to define rigorously. An intuitive way is to say that energy is the *capacity to do work*.

There are various classifications of energy, all having the dimension (in SI units) of kg m^2 s^{-2}.

2.1. Classification

If N_A is Avogadro's number, then $Nm = nN_A \, m = nM$, where M is the molar mass. Thus,

$$PV = \frac{1}{3} nM \langle c^2 \rangle$$

Equating this to the ideal gas law gives, for $n = 1$,

$$\sqrt{\langle c^2 \rangle} = \sqrt{3RT/M}$$

A. Kinetic and Potential Energy. Kinetic energy is the energy an object possesses by virtue of its motion. Potential energy is the energy an object possesses by virtue of its position or composition. This classification is not very useful in Thermodynamics.

B. In Thermodynamics the most useful forms of energy are Work, Heat, Internal Energy, and related forms, such as Enthalpy, Free Energy, etc. to be introduced later.

2.2. System and Surrounding

We now introduce concepts which are essential to the application of
thermodynamics, namely

- *system*, which is part of the universe in which one is interested, and
- *surrounding*, the rest of the universe or what is not the system.

It is convenient to characterize a system as either

- **Open**, which is a system that can exchange energy and matter with the
  surrounding, or
- **Closed**, which is a system that cannot exchange matter with surrounding, or
- **Isolated**, which is a system that cannot exchange energy or matter with
  the surrounding.

2.3. Work and Heat

Work and **Heat** are always associated with *transfer* of energy between
systems and surrounding. More explicitly, **Work** is the energy transfer
that can be used to move boundaries of a system, or lift weight. **Heat** is
the energy transfer due to temperature difference between the system and
surrounding. **Energy** is a property of the system.

> Comment: There are various types of work: work associated with
> expansion or compression (we will call that PV work), electrical work,
> gravitational and other types of work. We will denote the pressure–
> volume work as w_{PV} and all types of work as w_{other}. For the most
> part, we will be concerned in this course with PV work.

Temperature obviously plays an important role in thermodynamics. The
definition used here — thinking of temperature as a measure of degrees
of hotness or coolness — is obviously not rigorous. There is a rigorous
approach, the axiomatic *approach*, but it will not be pursued in this course.

From a *molecular* (as opposed to thermodynamic) point of view, work
is a form of energy transfer that utilizes *uniform* motion of the molecules
in the surrounding. Heat is energy transfer that involves *chaotic* motion
of the surrounding molecules.

2.4. Measurement of Work

Work is *distance times force*. **Force** is mass times acceleration.
The following are examples of work:

a) When lifting an object, a force must be applied in the direction away
from the earth, which is equal to the mass times the gravitational
acceleration, i.e. $m \times g$. If the object moves a distance h, the work
done by the surrounding on the system is

$$w = h \times mg \tag{2.1}$$

b) When a gas in a cylinder (see Fig. 2.1) is compressed, the piston, acted
on by the external force f_{ext}, moves a distance dx in the direction
of f_{ext}. The external pressure on the piston of area A is $P_{ext} = f_{ext}/A$. The piston is displaced a distance dx in the direction of the
force, f_{ext}.

The change in the volume of the system is $dV = -Adx$. (The
negative sign is introduced because the final volume V_f is smaller than
the initial volume V_i.) Accordingly, the work done by the surrounding on
the system is

$$dw_{PV} = (f_{ext}/A)Adx = -P_{ext}dV \tag{2.2}$$

For constant P_{ext} the integrated value is

$$w_{PV} = -P_{ext}\Delta V \tag{2.3}$$

where $\Delta V = V_f - V_i$. Obviously, the maximum work is obtained when the
internal gas pressure is infinitesimally less than the external. If greater,
compression cannot occur.

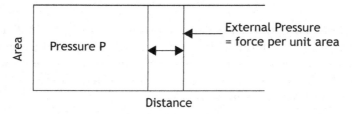

Fig. 2.1   Compression of a gas by piston.

2.5. Reversible Process

When the internal pressure P is equal to the external one P_{ext}, the system is said to be in a state of equilibrium. An infinitesimal increase in P_{ext} will result in an increase in compression and an infinitesimal decrease in P_{ext} will result in an increase in expansion. A ***reversible process*** is a process which proceeds through a ***succession of equilibrium states***.

If an ideal gas is in equilibrium with its surroundings, then $P_{\text{ext}} = P$ and, for an ideal gas, $P = nRT/V$. Therefore, the work (done by the surrounding on the system) in a ***reversible*** change is

$$w_{\text{PV}} = -\int (nRT/V)\mathrm{d}V = -nRT \ln V_{\text{f}}/V_{\text{i}} \tag{2.4}$$

Note again that the work is positive in a compression ($V_{\text{f}} < V_{\text{i}}$) and negative in an expansion ($V_{\text{f}} > V_{\text{i}}$).

2.6. Measurement of Heat

The traditional way to discuss the concepts of temperature and heat is to define one of these and deduce the other from it. Attempts to define temperature in terms of heat are bound to cause difficulties since normally heat is not observed directly but inferred from changes in temperature. Statements such as *radiant energy, thermal heat flow*, are sometimes used in defining heat. Using such descriptions of heat to define temperature is obviously not very satisfactory and will be avoided here.

> Comment: The concept of heat is most conveniently described in terms of the heat capacity, which is the heat divided by the temperature change (Section 10). Here, we introduce the concept by focusing directly on temperature changes of two systems in thermal equilibrium.

Consider a system, A, initially at a temperature T_{A} in equilibrium with a system, B, whose initial temperature is T_{B}. If the equilibrium temperature is T, small changes in temperature of the two systems can be written as $\mathrm{d}T_{\text{A}} = T - T_{\text{A}}$ and $\mathrm{d}T_{\text{B}} = T - T_{\text{B}}$. The ratio of these quantities defines the ratio of the ***heat capacities***, that is

$$\mathrm{d}T_{\text{A}}/\mathrm{d}T_{\text{B}} = -C_{\text{A}}/C_{\text{B}} \tag{2.5}$$

(The minus sign is introduced, because one of the dT's has to be negative and we require the C's to be positive.) If a particular value is assigned to one of the heat capacities, the other is automatically established.

Once heat capacity is defined, heat transferred can be expressed by the relation,

$$q/\Delta T = C \tag{2.6a}$$

or
$$dq = CdT \tag{2.6b}$$

For macroscopic systems,

$$q = \int CdT \tag{2.7}$$

and, if the heat capacity is constant, $q = C\Delta T$.

2.7. Internal Energy

When the surrounding does work on or supplies heat to the system, the surrounding loses energy. But energy cannot be lost or gained — energy must be conserved. The energy lost by the surrounding is gained by the system in the form of **internal energy**. [Note: since the advent of relativity, energy and matter can be interconverted, and a more accurate statement would be: Energy–matter must be conserved.]

Conservation of energy in Thermodynamics is effectively the **First Law of Thermodynamics**. It can be expressed as

$$\Delta U = q + w \tag{2.8}$$

where, in general, w stands for total work: $w = w_{PV} + w_{other}$.

Convention about signs

In this course, as indicated previously, (and used in most scientific but not engineering treatments,) *w represents the work done by the surrounding on the system and q is the heat supplied by the surrounding to the system.*

Comment: Consistent with this convention is the formula $dw_{PV} = -P_{ext}dV$, which tells us that w is positive in a compression and negative in an expansion. [Engineers often use the convention that work is positive when done by the system on the surrounding, and their First Law reads $\Delta U = q - w$.]

Note: U is a *state* function, meaning it is independent of the previous history of the system but depends only on the current state and not on the way the state was formed. The quantities q and w are not state functions.

There are other state functions, to be introduced later. All will be denoted by capital letters in contrast to the concepts of work and heat, which are denoted by small letters.

2.8. Exact and Inexact Differentials

The First Law is frequently expressed in differential form:

$$dU = dq + dw \tag{2.9}$$

There is a difference between the differentials dU on the one hand and dq and dw on the other hand. dU is an exact differential — its *integral depends only on the initial and final states of the system and not on the path of integration*. The differentials dq and dw are generally not exact. Their integrated values depend on the path of integration.

As an illustration of the meaning of exact and inexact differentials, consider the integration of ydx (the horizontally shaded area) between the limits A and B. Obviously, the value depends on the path of integration. The same is true for the integral xdy. But the sum of these two is independent of the path. This shows that individual integrals may be path-dependent, but their sum could be path-independent (see Fig. 2.2).

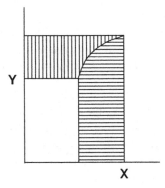

Fig. 2.2 Graphical representation of the sum of the integrals $\int ydx + \int xdy$.

2.9. Relation of ΔU to q_V (q at constant volume)

Consider an expansion against a constant external pressure P_{ext}. There is only PV work.

$$\Delta U = q - P_{ext}\,\Delta V \qquad (2.10)$$

$$dU = dq - P_{ext}\,dV \qquad (2.11)$$

If the volume is constant,

$$\Delta U = q_V \qquad (2.12)$$

$$dU = dq_V \qquad (2.13)$$

These are useful relations, because they allow the internal energy change to be obtained from measurements of q_V, say in a bomb calorimeter. However, chemists as a rule do not work with constant volumes. Question: Is there a state function which can be simply related to q_P, the heat at constant pressure? There is! That function is the **enthalpy**, H, defined as

$$H = U + PV \qquad (2.14)$$

(H is a state function, because U is and so are P and V.)

Consider a change at constant pressure P. Then,

$$\Delta H = \Delta U + P\Delta V \qquad (2.15)$$

If there is only PV work, then at constant pressure, with $P_{ext} = P$,

$$\Delta U = q_P + w_{PV} = q_P - P\Delta V \qquad (2.16)$$

and

$$\Delta H = q_P \qquad (2.17a)$$

$$dH = dq_P \qquad (2.17b)$$

It is important to keep in mind that the above relation is valid only if there is no work other than PV work. If there is other work, w_{other}, the change in H at constant temperature will be

$$\Delta H = q_P + w_{other} \qquad (2.18)$$

2.10. Heat Capacity

The heat capacity, as noted earlier, is the heat absorbed divided by the change in temperature. $C = q/\Delta T$ or, in differential form, $C = \mathrm{d}q/\mathrm{d}T$. The heat capacity depends on the amount of material. Most often used are the **molar heat capacity** (heat capacity per mole), C_{mol}, and specific **heat capacity** (heat capacity per gram), c_{g}. In scientific work, the use of molar heat capacity is standard.

In addition to characterizing the heat capacity by the amount of material, it is also necessary to specify the condition under which the change takes place. The heat capacities most often used are molar heat capacity at constant volume, $C_{\mathrm{V,m}}$, and molar heat capacity at constant pressure, $C_{\mathrm{P,m}}$. In the absence of work other than PV work, $\mathrm{d}U = \mathrm{d}q - P\mathrm{d}V$ and $\mathrm{d}H = \mathrm{d}U + P\mathrm{d}V + V\mathrm{d}P = \mathrm{d}q + V\mathrm{d}P$. Obviously, at constant volume, $\mathrm{d}q_{\mathrm{v}}/\mathrm{d}T = (\partial U/\partial T)_{\mathrm{V}} = C_{\mathrm{V}}$ and at constant pressure, $\mathrm{d}q_P/\mathrm{d}T = (\partial H/\partial T)_{\mathrm{P}} = C_{\mathrm{P}}$. It is pointed out that $\mathrm{d}q_P = \mathrm{d}U + P\mathrm{d}V$ [Eq. (2.16)].

For one mole of an ideal gas, $PV = RT$, and so

$$C_{\mathrm{P,m}} - C_{\mathrm{V,m}} = \mathrm{d}q_P/\mathrm{d}T - \mathrm{d}q_{\mathrm{v}}/\mathrm{d}T = P(\partial V/\partial T)_{\mathrm{P}} = PR/P = R \quad (2.19)$$

2.11. Enthalpy Changes in Chemical Reactions

What makes the enthalpy particularly useful is that the integral is independent of the path of integration. For example, if one wants to know the enthalpy change for

$$C + \frac{1}{2}O_2 \rightarrow CO \quad (2.20)$$

which is difficult to produce, one can obtain the result from the following reactions, which are easy to measure

$$C + O_2 \rightarrow CO_2; \qquad \Delta H = -94.1\,\mathrm{kcal\,mol}^{-1} \quad (2.21)$$

$$CO_2 \rightarrow CO + \frac{1}{2}O_2; \quad \Delta H = +67.6\,\mathrm{kcal\,mol}^{-1} \quad (2.22)$$

Adding the two reactions gives Eq. (2.20) with $\Delta H = -26.5\,\mathrm{kcal\,mol}^{-1}$.

2.12. Standard Enthalpy

Standard enthalpy, denoted as $H^{\underline{o}}$, is the enthalpy of a system at one bar pressure and a specified temperature (usually 25°C). Standard enthalpy of **formation** is the *standard enthalpy of reaction of the formation of a substance formed from its elements in their standard states*. The standard enthalpy of formation of an *element is taken to be zero*.

2.13. Variation of Enthalpy with Temperature

Returning to the concept of heat capacity, $C = q/\Delta T$, or more precisely

$$C = \lim_{\Delta T \to 0} (q/\Delta T) = dq/dT \tag{2.23}$$

we can write for C_P

$$(\partial H/\partial I)_P = C_P \text{ and } dH = C_P dT \tag{2.24}$$

If the temperature, originally at T, changes to T' and C_P is constant

$$H(T') = H(T) + C_P(T' - T) \tag{2.25}$$

For a chemical reaction,

$$\Delta_r C_P = \Sigma_i n_i C_{P,ni}(\text{products}) - \Sigma_i n_i C_{P,ni}(\text{reactants}) \tag{2.26}$$

$$\Delta_r H = \Sigma n_i H_i^O(\text{prod}) - \Sigma n_i H_i^O(\text{react}) \tag{2.27}$$

$$\Delta_r H^O(T') = \Delta H^O(T) + \Delta C_P \Delta T \tag{2.28}$$

Chapter 3

The Second Law of Thermodynamics

The First Law enables one to determine energy changes that accompany given processes. The First Law does not say whether such processes occur (spontaneously!) or do not occur. Such and many other experimental facts are not covered by the First Law.

For example,

1) When two bodies in thermal contact are brought together, heat will flow from the hotter to the colder body. The reverse process (heat flowing from the colder to the hotter body) does not even occur. There would be no violation of the First Law if heat would flow from the colder to the hotter body as long as there is no net gain or loss of energy.

2) Expansion of a gas into a vacuum occurs spontaneously, but the reverse does not occur.

3) Some chemical reactions occur spontaneously, others do not. It was once thought that if ΔU is negative (energy is lowered), the change will occur spontaneously. This is frequently true but not always. For example, the reaction H_2O (s, $10°C$) $\rightarrow$ H_2O (l, $10°C$) occurs spontaneously although ΔU is positive ($\sim +1.5$ kcal mol^{-1}).

4) It was once thought also that a negative ΔH is the proper criterion for spontaneity. This is generally the case but not always. For example, ΔH for the reaction $Ag(s) + 1/2Hg_2Cl_2(s) \rightarrow AgCl(s) + Hg(l)$ is positive ($+1.28$ kcal mol^{-1}) but occurs spontaneously.

5) Some solvents (for example, benzene and toluene) mix completely although *heat* has to be supplied and the *enthalpy change is positive*.

Apparently, the First Law does not account for a large number of observations. Evidently another Law is needed and that other Law is the **Second Law of Thermodynamics**.

Actually, the Second Law is not directly concerned with questions of spontaneity. Rather, a new concept (the entropy) is introduced, which as a consequence of the second Law, behaves in a characteristic manner, depending on whether the change is reversible or irreversible. This new function, **the entropy**, **S**, is defined as

$$dS = dq_{rev}/T \tag{3.1}$$

or if T is constant,

$$S = q_{rev}/T \tag{3.2}$$

It is important to keep in mind that the simple relation between S and q is valid only under reversible conditions. This does not mean that when the process is not reversible, entropy does not exist but rather that under those circumstances *entropy is not simply related to q*.

The entropy functions have the following characteristics:

1) S is a state function, i.e. the integral $\int dS$ is independent of path. This means that the integral of dq_{rev}/T is also independent of path in spite of the fact that dq_{rev} is path-dependent.
2) In an *isolated* system, any transformation will result in

$$\Delta S > 0 \quad \text{if change is irreversible or spontaneous} \tag{3.3}$$
$$\Delta S = 0 \quad \text{if change is reversible or in equilibrium} \tag{3.4}$$

Since all naturally occurring phenomena are irreversible changes, **the system, if isolated, experiences an increase in entropy.**

3) For a non-isolated system at constant temperature,

$$\Delta S = q_{rev}/T \quad \text{if the process proceeds reversibly} \tag{3.5}$$
$$\Delta S > q_{irr}/T \quad \text{if the process proceeds irreversibly} \tag{3.6}$$

3.1. Statements of the Second Law

There are several ways to express the Second Law (Laws are often called Principles), which are based on the observation that heat cannot spontaneously flow from a colder to a hotter body. In more fancy language, the statement is usually expressed in the form of two equivalent Principles:

Claussius Principle: It is not possible to construct an engine which has the sole effect of transferring heat from a lower temperature reservoir to a higher one.

or,

Kelvin–Planck Principle: It is not possible to construct an engine which has the sole effect of converting heat entirely into work.

It can be shown that either statement leads to the conclusion that $\int dq_{rev}/T$ is path-independent and this statement is used to define entropy. It can be shown also *rigorously* that statements (3.3)–(3.6) are valid.

3.2. Carnot Cycle

The standard way of proving the foregoing statements is by means of a Carnot cycle. This is a reversible cycle consisting of an isothermal expansion of a gas from a volume V_1 to a volume V_2 at a temperature T_2 , followed by an adiabatic (no heat transfer) expansion of the gas from V_2 to V_2', then followed by an isothermal compression from V_2' to V_1' at a temperature T_1, and finally followed by an adiabatic compression from V_1' to V_1.

The system (gas) is connected to an engine, consisting of two heat reservoirs, one at the higher temperature, T_2, and the other at the lower temperature, T_1 (see Figs. 3.1 and 3.2). The heat, q_2, extracted from the high temperature reservoir T_2 is used to run the Carnot cycle, thereby generating work, $-w$, and returning the remaining heat q_1 to the reservoir at the lower temperature T_1. [Note: q_2 and q_1 are symbolic representations: q_2 is positive, and q_1 is negative. Work done on the surrounding has a negative sign because we use the convention that w represents the work done by the surrounding on the system, and thus the work done by the system must be $-w$.]

3.3. Engine Efficiency

The efficiency, ξ, is defined as the work done by the system in a complete cycle divided by the heat absorbed at the higher temperature, i.e.

$$\xi = -w/q_2 \tag{3.7}$$

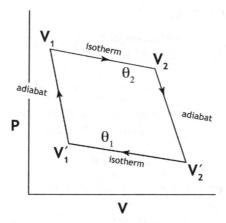

Fig. 3.1 Carnot cycle. The curves represent two isotherms and two adiabats. The upper isotherm is the higher temperature.

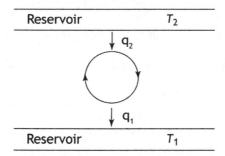

Fig. 3.2 Carnot cycle running in a clockwise direction, transferring heat from the high temperature reservoir to the lower.

The work done by the system on the surrounding must be represented by $-w$ because by our convention w represents the work done on the system.

Note that in the complete Carnot cycle, $\Delta U = 0$ (why?) and since $\Delta U = q + w$ ($q = q_2 + q_1$), it follows that $q + w = 0$. Thus,

$$\xi = (q_2 + q_1)/q_2 = 1 + q_1/q_2 \tag{3.8}$$

3.3.1. *Reversible Process*

Using an ideal gas as the work producing system, we can write for the high temperature *isothermal* transition

$$dU = 0 = dq_2 - PdV = dq_2 - RT_2 dV/V \qquad (3.9)$$

keeping in mind that an ideal gas varies only with temperature. Integrating from V_1 to V_2 at constant T_2 gives (see Fig. 3.1)

$$q_2 = RT_2 \ln V_2/V_1 \qquad (3.10a)$$

Similarly, the low temperature transition at T_1 gives

$$q_1 = RT_1 \ln V_1'/V_2' \qquad (3.10b)$$

Adiabatic changes of ideal gases obviously obey the rule

$$dq = dU - RT dV/V = 0 \qquad (3.11a)$$

Recall that at constant volume $\Delta U = q_V$ and so

$$(dU/dT)_V = dq_V/dT = C_V \text{ and therefore, } dU = C_V dT.$$

Accordingly, we can express (3.11a) as

$$dq = C_V dT + \frac{RT}{V} dV = 0 \qquad (3.11b)$$

Dividing by T, gives

$$C_V dT/T + R dV/V = 0 \qquad (3.11c)$$

Recalling that $C_P - C_V = R$ for an ideal gas, and introducing the symbol $\gamma = C_P/C_V$, we get, upon dividing by C_V,

$$dT/T + [R/C_V]dV/V = dT/T + [(C_P - C_V)/C_V]dV/V$$
$$= dT/T + (\gamma - 1)dV/V$$

Integration along the adiabat from the high isotherm (T_2, V_2) to the low isotherm (T_1, V_2') (see Fig. 3.1) gives

$$\ln(T_1/T_2) + (\gamma - 1)\ln(V_2'/V_2) = 0 \qquad (3.12a)$$

Similarly, for the adiabatic transition from the lower (T_1, V_1') to the higher isotherm (T_2, V_1) we obtain

$$\ln(T_2/T_1) + (\gamma - 1)\ln(V_1/V_1') = 0 \qquad (3.12b)$$

which is equivalent to

$$\ln(T_1/T_2) + (\gamma - 1)\ln(V_1/V_1') = 0 \qquad (3.12c)$$

Obviously, comparing (3.12a) with (3.12c) shows that $V_2'/V_2 = V_1'/V_1$ or $V_2'/V_1' = V_2/V_1$. Substituting this result in Eq. (3.10b), shows that

$$q_1 = -RT_1 \ln V_2/V_1 \qquad (3.12d)$$

Finally, substituting this expression in Eq. (3.8) and using (3.10a) yields

$$\xi = 1 - T_1/T_2 \qquad (3.13)$$

Note that since $\xi = -w/q_2 = (q_1 + q_2)/q_2 = 1 + q_1/q_2$ it follws that $q_1/q_2 = -T_1/T_2$ and thus, $q_1/T_1 + q_2/T_2 = 0$. Replacing this by the differential form $\mathrm{d}q_{\mathrm{rev}}/T$ (emphasizing the reversible nature of the process) yields, upon integration along a closed contour, $\int \mathrm{d}q_{\mathrm{rev}}/T = 0$, which is another indication that $\mathrm{d}q_{\mathrm{rev}}/T$ is an exact differential.

3.3.2. *Irreversible Process*

What is the efficiency when the process is irreversible?

Let us first assume that the efficiency is greater for the irreversible than for the reversible case. Denoting the irreversible efficiency as ξ^* and the reversible as ξ, and since $\xi^* > \xi$, we must have

$$-w^*/q_2^* > -w/q_2 \qquad (3.14)$$

Consider now two Carnot cycles, the starred and the non-starred one, coupled together (Fig. 3.3). Let us, for simplicity, adjust the engines so that $-w^* = -w$. Dividing both sides of (3.14) by $-w^*$, $1/q_2^* < 1/q_2$ and therefore, $q_2 < q_2^*$. Hence, since the w's are the same, $q_1^* > q_1$. This means that heat, in the amount of $q_1^* - q_1$, is transferred from the lower temperature reservoir to the higher temperature one, in violation of the Clausius Principle. So the efficiency of an irreversible change cannot be greater than that of a reversible change.

Can the efficiency of an irreversible change be less than that of a reversible change? Yes, there is no prohibition! Since $\xi = 1 - T_1/T_2$ for

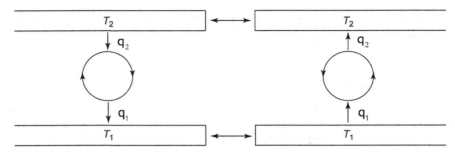

Fig. 3.3 Two coupled Carnot cycles. The one on the left runs in the forward direction; the one on the right runs in the reverse direction.

a reversible change, ξ^* must be less than $1 - T_1/T_2$ for an irreversible case. Hence, for the irreversible case, $(1 + q_1/q_2)_{irr} < (1 - T_1/T_2)$ which gives $q_1/T_1 + q_2/T_2 < 0$. In differential form this becomes, after integration along a closed contour,

$$\int dq_{irr}/T < 0 \qquad (3.15)$$

3.3.3. *General Changes in Entropy*

Consider a system undergoing a change from A to B irreversibly, and retuning from B to A (by another path) reversibly. Since part of the transformation is irreversible, the overall change is irreversible. Accordingly,

$$\int_A^B dq_{irr}/T + \int_B^A dq_{rev}/T < 0 \qquad (3.16)$$

Changing the second integration from A to B (which requires changing sign) gives

$$\int_A^B dq_{rev}/T > \int_A^B dq_{irr}/T \qquad (3.17)$$

Since the integral over the reversible heat represents the entropy change, we can say that in general,

$$\Delta S = S_B - S_A \geq \int_A^B dq/T \qquad (3.18)$$

where the sign $>$ refers to an irreversible change, and the $=$ sign refers to a reversible change.

3.3.4. *Isolated Systems*

Variation of heats is difficult to determine, and most often one calculates entropy changes of isolated systems, where $dq = 0$. Thus, for an isolated system, Eq. (3.18) takes the form

$$\Delta S_{\text{isolated}} \geq 0 \qquad (3.19)$$

Comment: The results obtained so far, which relates entropy to heat change divided by temperature, was derived for a particular system — an ideal gas. What assurance do we have that the efficiency criterion applies to other systems? It does! There is a theorem (not to be developed here) that proves that all systems operating *reversibly* between the same temperatures have the *same efficiency*. Thus, if we derive the entropy for one system, we have it for all.

3.4. Determination of Entropy

Note: We consider in this section only reversible changes.

3.4.1. *Entropy change in Phase Transitions (solid–liquid, liquid–vapor, solid–vapor)*

i) At constant T

$$\Delta S = \int dq/T = q/T \qquad (3.20)$$

ii) At constant T and P, $q_P = \Delta H$ and $\Delta S = \Delta H/T$ $\qquad (3.21)$

Example 3.1.

$$H_2O\,(s, 0°C) = H_2O\,(l, 0°C) \quad \Delta H = 6.01\,\text{kJ}\,\text{mol}^{-1}$$
$$\Delta S = 6010\text{J}\,\text{mol}^{-1}/273.15\,\text{K} = 21.99\text{J}\,\text{K}^{-1}\,\text{mol}^{-1}$$

3.4.2. *Entropy change in (Ideal) Gas Expansion*

i) At constant $T, \Delta U = 0$. Hence,

$$q = -w_{PV} = -\int (-P_{ext} dV) = \int P dV \qquad (3.22)$$

$$= nRT \int dV/V = nRT \ln V_f/V_i \qquad (3.23)$$

$$\Delta S = nR \ln V_f/V_i \qquad (3.24)$$

ii) For variable $T, \Delta S = \int dq/T = \int C dT/T$ and if C is independent of T, we have at constant pressure,

$$\Delta S = C_P \int dT/T = C_p \ln T_f/T_i \qquad (3.25)$$

and at constant volume

$$\Delta S = C_V \int dT/T = C_V \ln T_f/T_i \qquad (3.26)$$

Example 3.2.

a) One mole of an ideal gas expands at $T = 298$ K from 24.79 L to 49.58 L in a reversible process. Calculate the change in entropy.

Solution

$$\Delta S = R \ln V_f/V_i = 8.3145 \text{ J K}^{-1} \text{mol}^{-1} \times \ln 2$$

$$= 5.76 \text{J K}^{-1} \text{mol}^{-1}$$

b) One mole of an ideal gas expands at $T = 298$ K into a vacuum from an initial volume of 24.79 L to a final volume of 49.58 L. Calculate ΔS.

Solution

Here, $w = 0$ (since $P_{ext} = 0$) and so $q = 0$. One cannot use $dS = dq/T$ because q is not q_{rev}. However, we know that the initial and final volumes and temperatures are the same as in Part (a). The entropy, being a state function depends only on the initial and final states, which are known from Part (b). Thus the value of ΔS is the same as in Part (a).

Chapter 4

The Third Law of Thermodynamics

Up to now we could describe only differences in H, E, i.e. ΔH, ΔE, but not absolute values of H or E. In the case of entropy, on the other hand, we can describe the concept in absolute values. What makes this possible is the Third Law of Thermodynamics. There are several statements of the Third Law. The one we are using here is

The entropy of a perfect crystalline solid at absolute zero is zero.

Comments: Note the restriction *perfect*. Perfect means that the system is in perfect thermodynamic equilibrium. But how can you tell whether this is so experimentally at these very low temperatures? You can't.

What can be done is to measure the heat capacity over a temperature range around $0\,\text{K}$ and compare the results with the statistical entropy calculated by statistical mechanics, which assumes perfect thermodynamic equilibrium. If the entropies of the calorimetric measurements and the statistical values agree, the crystalline solid is "perfect". Otherwise, the solid is imperfect.

There are many systems which are not perfect: CO, N_2O, H_2O are examples. All cases, observed so far, can be explained or rationalized. For example, CO has a very small dipole moment. At absolute zero the dipoles should all point in the same direction. But that does not happen. As the solid cools, there are some "frozen-in" structures, with dipole moments pointing in opposite directions. The low temperature crystalline solid is then not in "perfect" thermodynamic equilibrium. Similar arguments have been presented for the other exceptions.

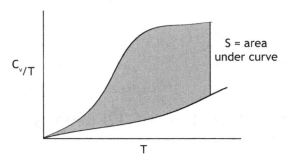

Fig. 4.1   Variation of S with T.

Still, because there are exceptions, some individuals are reluctant to call the Third Law a **Law** on par with the First and Second Laws. But most chemists accept the law, and use it as the basis for absolute entropies.

Suppose one wants to determine the entropy change $\Delta S = S(T) - S(0)$. Since $S(0)$ is assumed to be zero, one obtains

$$\Delta S(T) = S(T) = \int_0^T (C_P/T)\mathrm{d}T \tag{4.1}$$

A schematic representation of the variation of S with T is shown in Fig. 4.1.

4.1. Standard Entropy

Just as with standard enthalpy, $H^{\underline{o}}$, there is a standard entropy, $S^{\underline{o}}$, defined as the entropy of one mole of a substance at 1 bar and a specified temperature, normally 25°C.

4.2. Molecular Interpretation of Entropy

We have seen that when a system changes from a solid to a gas, from a solid to a liquid or from a liquid to a gas, there is an increase in entropy. We have also seen that in an irreversible isothermal change, as in the expansion into a vacuum of a gas at constant temperature, the entropy increases. Also, when there is an increase in temperature, the entropy increases.

Evidently, in all these cases, the increase in entropy seems to be accompanied from a microscopic (molecular) point of view by an increase in disorder, greater randomness in the distribution of the molecules, and more chaotic motion.

Example 4.1. Suppose you are dealing with a system consisting of particles, distributed among the energy levels shown in Fig. 4.2. The total

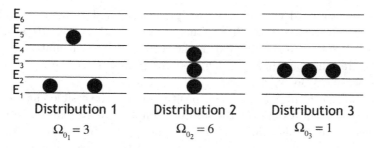

Fig. 4.2 Schematic diagram of the distribution of 3 particles among equally spaced energy levels.

energy of the three particles is 6 units. There are three different types of distributions, type D_1, type D_2, and type D_3. Type 1 can be realized in three ways; there are 3 microstates, $\Omega_{D1} = 3$, associated with this distribution. Type 2 can be realized in six ways; there are six microstates, $\Omega_{D2} = 6$. In type 3 there is one microstate, $\Omega_{D3} = 1$.

Had we used billions of particles instead of three, the most probable distribution would be so much more probable than all other distributions that, for all practical purposes, this would be the only distribution. Let us denote the number of microstates of this distribution as Ω_{D^*}.

Boltzmann thought that the more ways you can distribute the particles, the higher the entropy would be, and suggested that the entropy is related to the number of microstates by the formula

$$S = k \ln \Omega_{D^*} \tag{4.2}$$

where k is a constant (called Boltzmann's constant).

In a crystalline solid at $0\,\mathrm{K}$, all molecules are in their ground state and there is only one way to realize this, i.e. $\Omega_{D^*} = 0$, therefore, $S(0) = 0$.

A spontaneous process is a process in which the molecules can distribute themselves more randomly, either in terms of their energy distribution or their position. Hence, the entropy change in a spontaneous process, which is irreversible, is greater than the entropy change in a reversible process.

4.3. The Surroundings

The reaction $2H_2\ (g) + O_2\ (g) \rightarrow 2H_2O\ (l)$ proceeds spontaneously, in fact explosively, once initiated. Yet, ΔS is negative ($-327\ \mathrm{J\ K^{-1}\ mol^{-1}}$). How is that possible?

The standard entropies of H_2 (g), O_2 (g) and H_2O (l) at 25°C are respectively 130.7, 205.2, 70.0 kJ mol^{-1}, giving $\Delta S = (2 \times 70.0 - 205.2 - 2 \times 130.7)$ kJ $= -326.6$ kJ. [This is the correct way to calculate ΔS, since S is a state function. Had we calculated ΔS from q/T, the result would have been wrong. Why?]

The rule that ΔS has to be positive in an irreversible change applies to an *isolated* system. In the above example, the system is not isolated. To get around this difficulty, the standard procedure is to consider also the entropy of the surroundings (the rest of the universe), and regard the system and surroundings as an *isolated* system. Thus,

$$\Delta S_{\text{total}} = \Delta S_{\text{system}} + \Delta S_{\text{surroundings}} \qquad (4.3)$$

4.4. The Entropy of the Surroundings

In treating the surrounding entropy, it is common practice to assume that the surrounding changes *reversibly* even though the system may change irreversibly. [The rationale is that the surrounding, being so enormous, would change quasi-statically.]

Example 4.2. Let us return to the water reaction. It is found that at 25°C and one atmosphere the heat released is -572 kJ mol^{-1}. As stated, the heat released by the system is assumed to be absorbed by the surrounding reversibly even though the system looses heat irreversibly.

The heat absorbed by the surrounding is $q_P = +572$ kJ and thus $\Delta S_{\text{surr}} = 572$ kJ$/298$ K $= 1.92$ kJ K^{-1} $= 1920$ J K^{-1}. On the other hand, the system entropy, as noted before, is $\Delta S_{\text{sys}} = -327$ J K^{-1} and so ΔS_{tot} equals $(-327 + 1920)$ J K^{-1} $= 1539$ J K^{-1} and is positive, as it should be.

In general, at constant T, $\Delta S_{\text{surr}} = q_{\text{surr}}/T = -q_{\text{sys}}/T$ and if P is also constant, and if there is no work other than pV work,

$$\Delta S_{\text{surr}} = -\Delta H_{\text{sys}}/T \qquad (4.4)$$

Comment: ΔH can always be related to q, regardless of whether q is reversible or irreversible. ΔS can only be equated to q/T if q is reversible.

Chapter 5

The Free Energy Functions

There is a drawback in working with entropy except for isolated systems. If the system is not isolated, the system and surroundings have to be combined, and at constant T and P, using Eq. (4.4) ($\Delta S_{\text{surr}} = -\Delta H_{\text{sys}}/T$) yields

$$\Delta S_{\text{tot}} = \Delta S_{\text{sys}} + \Delta S_{\text{surr}} = \Delta S_{\text{sys}} - \Delta H/T \qquad (5.1)$$

This requires that

$$\Delta S_{\text{tot}} > 0 \text{ if process is spontaneous;} \qquad (5.2a)$$
$$= 0 \text{ if process is reversible;} \qquad (5.2b)$$
$$< 0 \text{ if process is impossible.} \qquad (5.2c)$$

Or, for short,

$$\Delta S_{\text{tot}} \geq 0 \qquad (5.2d)$$

There is a way to get around the problem of having to use an *isolated system* or a *system plus surrounding*, rather than focusing on the system only. This can be accomplished by using *free energy functions*, of which there are two kinds: *the Helmholtz free energy* and *the Gibbs free energy* (also called *Gibbs energy*). In this course, only the Gibbs function will be used.

5.1. The Gibbs Free Energy

The Gibbs free energy is a property of the system only (surroundings not included). The Gibbs energy is defined as

$$G = H - TS \qquad (5.3)$$

Comment: It should be emphasized that G is a *state* function, because H and S are state functions, meaning that the integral, $\Delta G = \int_A^B dG = G(B) - G(A)$, depends only on the initial and final states A and B.

At constant temperature,

$$\Delta G = \Delta H - T\Delta S \qquad (5.4)$$

or

$$-\Delta G/T = \Delta S - \Delta H/T \qquad (5.5)$$

and, if the pressure is also constant, we can write, using Eq. (4.4),

$$-\Delta G/T = \Delta S_{sys} + \Delta S_{surr} = \Delta S_{tot} \qquad (5.6)$$

Thus, instead of using entropy of the system and surroundings, we can use the Gibbs free energy, which refers to the system only, and write

$$\Delta G_{T,P}/T < 0 \text{ if process is spontaneous;} \qquad (5.7a)$$

$$\Delta G_{T,P}/T = 0 \text{ if process is reversible;} \qquad (5.7b)$$

$$\Delta G_{T,P}/T > 0 \text{ if process is impossible.} \qquad (5.7c)$$

Or for short,

$$\Delta G_{T,P} \leq 0. \qquad (5.7d)$$

Thus, if there is only PV work, $\Delta G_{T,P} = 0$ for a *reversible* change (system in equilibrium) and negative for an irreversible change. This is a powerful criterion for spontaneity.

The Gibbs energy discussed so far, although more convenient to work with than entropy, is still not the most general. It is based on the assumption that only PV work is present. If w_{other} is also present, then at constant P and T,

$$\Delta H = \Delta U + P\Delta V \qquad (5.8a)$$

$$= q_P - P\Delta V + w_{other} + P\Delta V \qquad (5.8b)$$

$$= q_P + w_{other} \qquad (5.8c)$$

When the system exchanges heat with the surroundings, the heat lost (gained) by the system is equal to the heat gained (lost) by the surroundings. The process is treated as reversible, i.e. $q_{sys} = -q_{surr} = -T\Delta S_{surr}$. Thus,

$$\Delta H = -T\Delta S_{surr} + w_{other} \qquad (5.8d)$$

For fixed T and P,

$$\Delta G_{T,P} = \Delta H - T\Delta S_{sys} = -T\Delta S_{surr} - T\Delta S_{sys} + w_{other}$$
$$= -T\Delta S_{tot} + w_{other} \qquad (5.9)$$

Since $\Delta S_{tot} \geq 0$, the right-hand side of Eq. (5.9) is less than or equal to w_{other} and

$$\Delta G_{T,P} \leq w_{other} \qquad (5.10)$$

Again, for the change to proceed reversibly (in equilibrium), the change in G at constant P and T has to be equal to the non-PV work, w_{other}. If irreversible, the change has to be less.

The relation between $\Delta G_{P,T}$ and w_{other} is very useful, as it enables us to obtain ΔG from work measurements. An example (to be discussed later) is the determination of the free energy from work measurements in electrochemical studies.

Comment: The Helmholtz free energy is defined as $A = U - TS$, where U and S are properties of the system only. This function is useful when the volume is constant, and, as indicated above, it will not be pursued here.

5.2. Free Energy Changes in Chemical Reactions

It is important to distinguish between reaction free energies under standard conditions, indicated by the superscript $^{\circ}$, and under all *other* conditions. The reason, as will be shown later, is that the *equilibrium constant K is related to the standard free energy per mole*, ΔG_m°, *and not to the actual free energy* ΔG.

Comments: It should be noted that the quantities ΔG_f, ΔH_f, etc. in a typical handbook, refer to the molar free energy, molar enthalpy, etc. and thus the dimensions are the dimensions of energy or Joules per mole. Any combination with stoichiometric coefficients, as in the reaction free energies or enthalpies, will produce values of units J/mol. To obtain the real dimensions of free energies or enthalpies, one should use mole numbers, n_i, instead of stoichiometric numbers, such as ν_i. Considering that $n_i = \nu_i \, mol^0$, where mol^0 is the unit mol, it follows that to obtain energy dimensions in J, all that is needed is multiplication of the final molar quantities by the unit mol.

The standard free energy can be obtained from the standard enthalpy of formation, $\Delta H_f^{\underline{o}}$, and the standard entropy, $S^{\underline{o}}$, or from the standard free energy of *formation*, $\Delta G_f^{\underline{o}}$, listed in many handbooks.

Example 5.1. Find $\Delta G^{\underline{o}}$ for the reaction

$$3O_2(g) \rightarrow 2O_3(g) \text{ at } 298.15\,K. \tag{5.11}$$

1) From $\Delta H^{\underline{o}}$ and $S^{\underline{o}}$,

$$\Delta H^{\underline{o}} = 2\Delta H_f^{\underline{o}}(O_3) - 3\Delta H_f^{\underline{o}}(O_2) \tag{5.12a}$$
$$= 2 \times 142.7 - 3 \times 0 = 285.4\,kJ$$
$$\Delta S^{\underline{o}} = 2S^{\underline{o}}(O_3) - 3S^{\underline{o}}(O_2) \tag{5.12b}$$
$$= 2 \times 238.93 - 3 \times 205.14 = -137.56\,JK^{-1}$$
$$\Delta G^{\underline{o}} = \Delta H^{\underline{o}} - T\Delta S^{\underline{o}} \tag{5.12c}$$
$$= 285.4\,kJ + 298.15 \times 137.56/1000\,kJ$$
$$= (285.4 + 41.0)\,kJ = 326.4\,kJ$$

2) From $\Delta G_f^{\underline{o}}$

$$\Delta G_f^{\underline{o}} = 2\Delta G_f^{\underline{o}}(O_3) - 3\Delta G_f^{\underline{o}}(O_2) \tag{5.13}$$
$$= (2 \times 163.2 - 0) = 326.4\,kJ$$

Note that the sign of ΔG is positive, and according to what was said before, the reaction as written cannot proceed spontaneously. The reverse reaction, $O_3 \rightarrow O_2$, on the other hand, has a negative ΔG and can proceed spontaneously. This is indeed happening in the upper atmosphere, where ozone is converted to oxygen.

5.3. Variation of *G* with *T* and *P*

From $G = H - TS$, we obtain at constant T

$$dG = dH - TdS \tag{5.14a}$$
$$= dU + PdV + VdP - TdS \tag{5.14b}$$

and if only PV work is present,

$$dG = dq - P_{\text{ext}}dV + PdV + VdP - TdS \tag{5.15}$$

for a reversible change (P being the system pressure, $P_{\text{ext}} = P$), we get

$$dG = TdS - PdV + PdV + VdP - TdS \tag{5.16}$$

Thus, at constant T, for reversible change and PV work only,

$$dG = VdP \tag{5.17}$$

For an ideal gas,

$$dG = (nRT/P)dP \tag{5.18}$$

and

$$\Delta G = nRT \int_{\text{i}}^{\text{f}} dp/P = nRT \ln P_{\text{f}}/P_{\text{i}} \tag{5.19}$$

If we define $P^{\underline{o}}$ as a standard pressure (say 1 atm or 1 bar at 298.15 K) corresponding to the standard free energy, $G^{\underline{o}}$, we can write

$$\Delta G = G - G^{\underline{o}} = nRT \ln P/P^{\underline{o}} \tag{5.20}$$

Comment: This expression can be used only if the gas is ideal. If the gas is not ideal, the pressure is often replaced with an *effective pressure*, called *fugacity*.

5.4. Generalization of the Free Energy. Activity

The form (but not the actual values) of the ideal gas formulas is also used in the treatment of other types of materials, including mixtures. This is accomplished by introducing a new concept, the activity. The activity, a_i, of species, i, is defined as

$$G_i = G_i^{\ominus} + n_i RT \ln a_i \qquad (5.21)$$

Note that the activity is defined in terms of G, which is a true value of the system. However, to obtain the exact value of a_i, one must know the value of G_i, which is generally not known exactly. What is done often is to approximate the activity as follows:

$$\text{For an (ideal) gas } a_i = P_i/P^{\ominus} \qquad (5.22a)$$

$$\text{For a pure liquid or solid } a_i = 1 \qquad (5.22b)$$

$$\text{For a solution } a_i = c_i/c^{\ominus} \qquad (5.22c)$$

The symbol $P^{\ominus}$ stands for standard pressure (760 Torr or 1 atm or 101.2325 kPa); c_i denotes the concentration of species i, and $c^{\ominus}$ represents the standard concentration (1 mol dm^{-3} or 1 kg m^{-3}).

5.5. Partial Molar, Molal Quantities

Suppose we have a mixture of n_A moles of pure A and n_B moles of pure B. Denoting *the molar volumes* of A and B respectively as V_A^*, and V_B^*, then in general, the total volume will not be the sum of the individual values, i.e. $V \neq V_A^*$, $+ V_B^*$. [This applies also to the other thermodynamic functions, such as H, S, G, etc.] The reason why this is so is that molecules attract or repel each other, causing deviations from the sum rule. In general, the volume of a mixture depends on temperature, pressure, mole fractions, etc.

It can be shown that if a system is homogeneous, the total volume is

$$V = \Sigma_i n_i V_{i,m} \qquad (5.23a)$$

and, in particular, in a two-component system,

$$V = n_A V_{A,m} + n_B V_{B,m} \qquad (5.23b)$$

The subscript m refers to molar quantity.

The Gibbs free energy can be written conveniently as

$$G = G(T, P, n_1, n_2, \ldots) \qquad (5.24)$$

The *partial molar* volume of species i is defined as the derivative of V with respect to n_i holding constant P, T, and all n_j not equal to n_i. Denoting the partial molar or molal volume of i as

$$V_i^* = (\partial V / \partial n_i)_{T, P n_j \neq n_i} \tag{5.25}$$

The total volume of all species of a homogeneous system can be written as

$$V = \Sigma_i n_i V_i^* \tag{5.26a}$$

and in particular in a two-component system,

$$V = n_A V_A^*, + n_B V_B^*, \tag{5.26b}$$

Similar considerations apply to the Gibbs free energy

$$G = n_A G_A^*, + n_B G_B^*, \tag{5.27}$$

5.6. The Chemical Potential

The partial molar free energy is often called the chemical potential and is denoted as

$$\mu_i = (\partial G / \partial n_i)_{T, P, n_j \neq n_i} \tag{5.28a}$$

Comment: The partial Gibbs free energy transcends ordinary properties of partials. The chemical potential, μ_i , can be defined also as the partial Helmholtz free energy, the partial internal energy, as well as the partial enthalpy, namely,

$$\mu_i = (\partial A / \partial n_i)_{T, V, n_j \neq n_i} \tag{5.28b}$$

$$\mu_i = (\partial U / \partial n_i)_{S, V, n_j \neq n_i} \tag{5.28c}$$

$$\mu_i = (\partial H / \partial n_i)_{S, P, n_j \neq n_i} \tag{5.28d}$$

As can easily be shown (not in this course), they are all equal to Eq. (5.28a). Thus, it makes sense to give these partials a special name; the name is *chemical potential*. Thus, instead of writing the Gibbs free energy in terms of the partial molar quantities, as in Eq. (6.10), it is common practice to express the free energy of, say, a two-component system, as

$$G = n_A \mu_A + n_B \mu_B \tag{5.29}$$

Comment: It should be emphasized that the chemical potential is an *intensive* property, which depends on T and P as well as on the composition but not on the amount.

We have shown in the last section that for one mole of *a one-component* system at pressure P,

$$G^* = G^{\underline{o}*} + RT \ln P_f/P^{\underline{o}} \tag{5.30}$$

In terms of activities [Eq. (5.27)]

$$G^* = G^{\underline{o}*} + RT \ln a \tag{5.31}$$

or

$$\mu = \mu^{\underline{o}} + RT \ln a \tag{5.32}$$

In a mixture, we can write for component i,

$$\mu_i = \mu_i^{\underline{o}} + RT \ln a_i \tag{5.33}$$

5.7. Relation of $\Delta G^{\underline{o}}$ to the Equilibrium Constant, K

Consider the reaction,

$$aA + bB \rightleftarrows cC + dD \tag{5.34}$$

Denoting the molar quantities of A as m_A^* and of B as m_B^*, we can express the free energy change of the reaction as

$$\Delta_r G = \{cG_C^{\underline{o}} + dG_D^{\underline{o}} - aG_A^{\underline{o}} - bG_B^{\underline{o}}$$
$$+ RT[c \ln a_C + d \ln a_D - a \ln a_A - b \ln a_B]\} \tag{5.35a}$$
$$= \Delta G^{\underline{o}} + RT \ln(a_C^c a_D^d / a_A^a a_B^b) \tag{5.35b}$$
$$= \Delta G^{\underline{o}} + RT \ln K \tag{5.35c}$$

At equilibrium, $\Delta G = 0$ and so

$$\ln K = -\Delta G^{\underline{o}} \tag{5.36a}$$

or

$$K = \exp(-\Delta G^{\circ}/RT) \tag{5.36b}$$

Comment: It is important to note that the equilibrium constant is not related to the actual free energy change (which is zero), but to the free energy change in the *standard state*.

5.8. Variation of K with T

Assuming that ΔH° and ΔS° vary negligibly with temperature, we have: at a constant temperature T

$$\ln K = -\Delta G^{\circ}/RT = -\Delta H^{\circ}/RT + \Delta S^{\circ}/R \tag{5.37a}$$

and at a different constant temperature T'

$$\ln K' = -\Delta G^{\circ}/RT' = -\Delta H^{\circ}/RT' + \Delta S^{\circ}/R \tag{5.37b}$$

resulting in

$$\ln K' = \ln K + \Delta H^{\circ}/R(1/T - 1/T') \tag{5.38}$$

Example 5.2. An ideal gas, initially at $T = 273\,\mathrm{K}$, $P = 1\,\mathrm{atm}$ and $V_{\mathrm{i}} = 22.4\,\mathrm{L}$ expands isothermally and reversibly to a final volume, $V_{\mathrm{f}} = 44.8\,\mathrm{L}$.

a) Calculate $w, q, \Delta U, \Delta H, \Delta S, \Delta G$.

The process is reversible so that

$$w = -\int P_{\mathrm{ext}}\mathrm{d}V = -nRT \int \mathrm{d}V/V = -RT \ln V_{\mathrm{f}}/V_{\mathrm{i}}$$
$$= -RT \ln(44.8\,\mathrm{L}/22.4\,\mathrm{L}) = -8.3145\,\mathrm{JK^{-1}mol^{-1}} \times 273\,\mathrm{K} \times \ln 2$$
$$= -1.573\,\mathrm{kJ}$$

$\Delta U = 0$ (ideal gas at constant temperature). Thus, $q = -w$ and so

$$q = 1.573\,\mathrm{kJ}$$
$$\Delta H = \Delta U + \Delta(PV) = 0 + \Delta(RT) = 0$$
$$\Delta S = q_{\mathrm{rev}}/T = 1,573.3\,\mathrm{J}/273\,\mathrm{K} = 5.763\,\mathrm{JK^{-1}}$$
$$\Delta G = \Delta H - T\Delta S = 0 - 273\,\mathrm{K} \times 5.763\,\mathrm{JK^{-1}} = -1.573\,\mathrm{kJ}$$

b) The gas expands into a vacuum from the same V_i to the same V_f as in Part (a). Calculate $w, q, \Delta U, \Delta H, \Delta S, \Delta G$.

$$w = 0 \text{ because } P_{\text{ext}} = 0 \text{ and thus } q = 0.$$

The other functions are state functions with the same initial and final states as in Part (a) and so the Δ values must be the same as in Part (a). In other words,

$$\Delta U = 0, \Delta H = 0, \Delta S = 5.763 \,\text{kJ and } \Delta G = -1.573 \,\text{kJ}.$$

Chapter 6

Phase and Chemical Equilibria

Two important applications of thermodynamics are phase equilibria and chemical equilibria.

6.1. Phase Equilibrium

We will treat here phase equilibria involving one-component systems and two-component systems. No chemical reactions are assumed to occur. In the next section, we will take up chemical reactions.

Figure 6.1 depicts a phase diagram of a one-component system, namely a plot of P vs. T. There are three areas and three lines. The areas represent the phases gas, liquid and solid. The lines represent regions of coexistence of gas–liquid, liquid–solid and solid–gas phases. The point where the lines meet is called the triple point.

6.1.1. *The Phase Rule*

The Phase Rule tells how many independent intensive variables (pressure, temperature, mole fraction, etc.) there are that can be varied in each region. These variables are normally referred to as *degrees of freedom*. The phase rule reads

$$F = c - p + 2, \tag{6.1}$$

Here, F stands for the variance or degrees of freedom; c represents the number of components and p the number of phases.

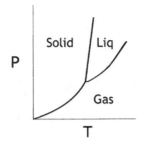

Fig. 6.1 Pressure–temperature phase diagram of a one-component system.

For example, in a one-component system,

$$F = 1 - 1 + 2 = 2 \quad \text{when } p = 1 \text{ (in the areas)}$$

$$F = 1 - 2 + 2 = 1 \quad \text{when } p = 2 \text{ (along the lines)}$$

$$F = 1 - 3 + 2 = 0 \quad \text{when } p = 3 \text{ (at the triple point)}$$

The phase rule tells us that in areas you can choose both the temperature and pressure. Along lines you can specify either the temperature or pressure but not both. At the triple point you cannot choose any variable: the variables are fixed.

6.1.2. *The Clapeyron Equation*

The slopes of the phase diagram lines can also be predicted thermodynamically, and must obey the *Clapeyron equation*:

$$dP/dT = \Delta H_{\text{trans}}/(T \Delta V_{\text{trans}}) \tag{6.2}$$

where the subscript trans refers to the particular transition, say liquid–vapor or solid–liquid.

The Clapeyron equation is easy to derive. From

$$G = H - TS$$

we get the differential form

$$dG = dH - T dS - S dT \tag{6.3a}$$

$$= dU + P dV + V dP - T dS - S dT \tag{6.3b}$$

and, for a reversible transition with PV work only,

$$dG = dq - PdV + PdV + VdP - TdS - SdT \tag{6.3c}$$

$$= TdS + VdP - TdS - SdT \tag{6.3d}$$

$$= -SdT + VdP \tag{6.3e}$$

Note: This is a generalization of an expression derived earlier where P was constant.

Consider two phases, α and β, in equilibrium at temperature T. Then, $G^\alpha = G^\beta$. Now consider an infinitesimal change from T to $T + dT$. The G's will then change to $G^\alpha + dG^\alpha$ and $G^\beta + dG^\beta$. If at the new temperature, $T + dT$, the systems are again in equilibrium, we must have $G^\alpha + dG^\alpha = G^\beta + dG^\beta$ and so

$$dG^\alpha = dG^\beta \tag{6.4a}$$

$$-S^\alpha dT + V^\alpha dP = -S^\beta dT + V^\beta dP \tag{6.4b}$$

which yields

$$(V^\alpha - V^\beta)dP = (S^\alpha - S^\beta)dT$$

$$\Delta V dP = \Delta S dT \tag{6.4c}$$

and thus,

$$dP/dT = \Delta S/\Delta V$$

$$= \Delta H/(T\Delta V) \tag{6.4d}$$

If a transition occurs between a liquid or solid and the vapor, then in general, the volume of the condensed system will be very much smaller than the volume of the gas. Neglecting the volume of the liquid, and treating the vapor as an ideal gas, the following are obtained:

liquid–gas $\quad dP/dT = \Delta H_{vap}/TV_{vap} \approx \Delta H_{vap}/(RT^2/P)$ $\quad$ (6.5a)

$dlnP/dT = \Delta H_{vap}/(RT^2)$ $\quad$ (6.5b)

solid–vapour $\quad dP/dT = \Delta H_{subl}/(TV_{vap})$ $\quad$ (6.6)

solid–liquid $\quad dP/dT = \Delta H_{fus}/[T(V_{liq} - V_{sol})]$ $\quad$ (6.7)

The subscripts vap, liq, subl, and fus refer respectively to vapor, liquid, sublimation and fusion. Equation (6.5b) is also known as the Claussius–Clapeyron equation.

Note: $\Delta H_{subl} > \Delta H_{vap}$ and, in general, $V_{liq} > V_{sol}$. (H_2O is an exception.)

Problem: Explain why, in general,

$$(dP/dT)_{liq-gas} < (dP/dT)_{sol-liq} \quad \text{and} \quad (dP/dT)_{sol-liq} > 0.$$

6.2. Chemical Equilibrium. Mixtures

There are three ways to characterize mole fractions:

(1) Molar concentration, $[x] = n_x/V_{solution}$ (x is the solute)   (6.8)
(2) Molal concentration, $m_x = n_x/M_{solvent}$ (usually mole/kg of   (6.9)
    solvent)
(3) Mole fraction, x_A/n (n = total number of moles)   (6.10)

6.2.1. *Ideal Solutions. Raoult's Law*

Let us adopt the convention that the *solvent* will be denoted as A and the *solute* as B.

The solvent of all substances, when sufficiently dilute, obeys Raoult's Law, that is

$$P_A = x_A \ P_A^* \tag{6.11}$$

where P_A^* is the vapor pressure of pure A.

The reason for this behavior can be rationalized that in very dilute solutions, a molecule A is surrounded essentially by other A molecules, but at a concentration of x_A. Hence, $P_A = x_A P_A^*$.

An Ideal Solution is a solution in *which every component obeys Raoult's Law over the entire range of compositions.*

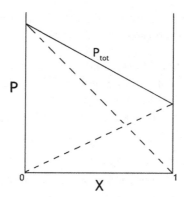

Fig. 6.2 Schematic diagram of an ideal solution. The dashed lines are the partial pressures of components A and B. The drawn line is the total pressure.

An example of an ideal solution is a mixture of benzene and toluene (see Fig. 6.2).

Note: Each partial pressure is $x_A \, P_A^*$ or $x_B \, P_B^*$ and the total pressure is the sum of the two.

It can be shown that when a substance obeys Raoult's Law, the chemical potential becomes

$$\mu_A = \mu_A^{\circ} + RT \ln x_A \qquad (6.12)$$

and thus $a_A = x_A$ [see also Eq. (5.33)].

6.2.2. *Ideal Dilute Solutions. Henry's Law*

Ideal solutions are not very common. Most solutions will show deviations from ideality. However, when solutions are very dilute, the solvent will obey Raoult's Law (as noted) and the *solute will obey Henry's Law.*

Henry's Law states that $P_B = x_B K_B$, where K_B is a constant, Henry's Law constant, generally not equal to the pressure of pure B, P_B^*. Henry's Law can also be defined in terms of the molarity $P_B = m_B K_H$. A solution in which the solvent obeys Raoult's Law and the solute obeys Henry's Law is called an *ideal dilute solution*. If Henry's Law constant is equal to P_B^*, then the ideal dilute solution becomes ideal.

Example 6.1. Methane (CH_4) and benzene (C_6H_6) form an ideal dilute solution. The pressure of pure benzene is $P_{benz} = 300\,\text{Torr}$ at $298\,\text{K}$. Henry's Law constant of the solute (methane) is $K_{meth} = 4.27 \times 10^4\,\text{Torr}$ at $298\,\text{K}$. If the mole fraction of CH_4 in the liquid phase is $x_{meth} = 1.01 \times 10^{-2}$, find

(a) the partial pressure of CH_4 at $298\,\text{K}$;
(b) the partial pressure of C_6H_6;
(c) the mole fraction of CH_4 in the gaseous phase.

Solution

(a) By Henry's Law, $P_{meth} = x_{meth}K_{meth} = 1.01 \times 10^{-2} \times 4.27 \times 10^4\,\text{Torr} =$ 431.27 Torr
(b) The solvent obeys Raoult's Law

$$P_{benz} = x_{benz}P_{benz}^o = (1 - 1.01 \times 10^{-2}) \times 300\,\text{Torr} = 296.97\,\text{Torr}$$

(c) Assuming ideal behavior in the gaseous phase,

$$x_{meth} = P_{meth}/(P_{met} + P_{benz}) = 431.27/(431.27 + 296.97) = 0.529$$

Example 6.2. Predict whether natural water can support life. It is known that to support aquatic life, the concentration of O_2 must be $4\,\text{mg/L}$. What must the partial pressure of O_2 in air be to achieve that concentration?

In a liter of water, there are 55.5 moles of H_2O and a negligible amount of O_2. The O_2 molar fraction is

$$x_{O2} = (4 \times 10^{-3}\,\text{g}\,\text{L}^{-1}/32.00\,\text{g}\,\text{mol}^{-1})/55.5\,\text{mol}\,\text{L}^{-1}$$
$$= 2 \times 10^{-6} \tag{6.13}$$

Henry's Law constant for O_2 is 3.79×10^7 Torr in water at $25°C$. Thus, $P_{O2} = x_{O2}K_{O2} = 2 \times 10^{-6} \times 3.60 \times 10^7\,\text{Torr} = 70\,\text{Torr}$. This is the minimum pressure O_2 must have in the air. The actual pressure of O_2 in the air is about 21% of 1 atm or $0.21 \times 760\,\text{Torr} = 160\,\text{Torr}$. Thus, there is more than ample O_2 in the air to sustain life in water on earth.

6.2.3. *Colligative Properties*

There are several properties, all resulting from *lowering the pressure of the solvent* that can be observed in solutions. They are

- Elevation of boiling point;
- Depression of freezing point;
- Production of osmotic pressure.

In the simplest cases, one assumes that

(1) the solute is not volatile;
(2) the solute does not precipitate.

Thus, when a water solution freezes, the frozen solid is pure ice.

6.2.4. *Elevation of Boiling Point. Depression of Freezing Point*

The easiest way to ascertain why there is a boiling point increase or a freezing point decrease is to observe the variation of the chemical potential with temperature.

Figure 6.3 shows three curves: the chemical potential of pure A in the vapor phase, pure A in the liquid phase and of A in the solvent. Notice that the solvent curve is lower than the liquid phase curve, as it should be because of the lower chemical potential. The pure vapor phase curve cuts the pure liquid curve at a lower temperature than the solution phase, indicating that the boiling point of the solution is higher.

Figure 6.4 shows the variation of the chemical potentials with temperature of pure A in the liquid phase, pure A in the solid phase and A in solution. Again the solution curve is lower than the liquid curve, as it should be. The pure liquid curve cuts the solution curve at a lower temperature than the pure liquid curve, indicating that the solution freezing point is lower.

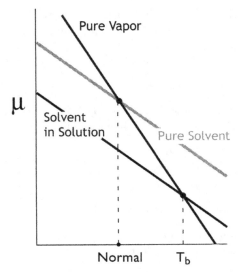

Fig. 6.3 Schematic diagram of the chemical potentials responsible for the elevation of the boiling point.

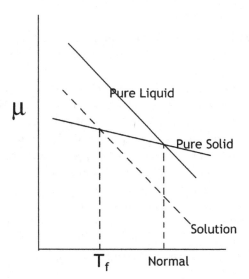

Fig. 6.4 Schematic diagram of the chemical potentials responsible for the depression of the freezing point.

Comment: The way the chemical potential behaves is similar to the behavior of the free energy. Thus, we have $\Delta\mu \leq 0$, where the equal sign pertains to equilibrium and the unequal sign to non-equilibrium situations. If a system can exist in two different states, the most favored state is the one with the *lowest* μ. When liquid and vapor are at equilibrium (coexist), say at the normal boiling point, T_b^*, the chemical potentials of the two phases must be equal. Above the normal boiling temperature, the vapor phase is the more stable state, and the chemical potential of the vapor will be lower than that of the pure liquid. Below the normal boiling temperature, the liquid phase is the stable phase, and the chemical potential of the liquid will be the lowest. Similar considerations apply to the chemical potentials responsible for the lowering of the freezing point (see Fig. 6.4).

Let us denote the *normal* boiling point and *normal* freezing point of the pure liquid respectively as T_b^* and T_f^* and the boiling and freezing points of the actual solutions as T_b and T_f. We can write

$$\Delta T_b = T_b - T_b^* = m_b K_b \tag{6.14a}$$

$$\Delta T_f = T_f^* - T_f = m_f K_f \tag{6.14b}$$

where m_b is the molality of solute and K_b and K_f are respectively the boiling point elevation constant and freezing point depression constant. (These constants can readily be related to thermodynamic functions, but will not be developed in this course.)

Example 6.3. Calculate the boiling point of a solution of $250\,\mathrm{cm}^3$ of $H_2O(l)$ containing $7.5\,\mathrm{g}$ of a solute. The normal boiling point of $H_2O(l)$ is $373.25\,\mathrm{K}$. The boiling-point elevation constant, K_b is $0.51\,\mathrm{K\,kg\,mol^{-1}}$. The molecular weight of the solute is $M = 342\,\mathrm{g\,mol^{-1}}$.

Solution
$7.5\,\mathrm{g}$ in $250\,\mathrm{cm}^3$ of water is equivalent to $30.0\,\mathrm{g}$ of solute in $1\,\mathrm{L}$ of water. The mole fraction of $30.0\,\mathrm{g}$ of solute in $1\,\mathrm{L}$ of $H_2O(l)$ is $m = 30.0\,\mathrm{g}/342\,\mathrm{g\,mol^{-1}} = 8.77 \times 10^{-2}\,\mathrm{mol}$ and thus the molarity of the solute is $8.77 \times 10^{-2}\,\mathrm{mol/L}$. Assuming that $1\,\mathrm{L}$ of water at the normal boiling point weighs essentially $1\,\mathrm{kg}$, we can equate the molarity m to the molality, m_b, and write $m_b = $

8.77×10^{-2} mol/kg. The elevation of the boiling point is $\Delta T_b = T_b - 373.25\,\text{K} = m_b K_b$. Thus,

$$T_b = 373.25\,\text{K} + 8.77 \times 10^{-2}\text{mol/kg} \times 0.51\,\text{K kg mol}^{-1}$$
$$= (373.25 + 0.0447)\text{K} = 373.29\,\text{K}.$$

6.2.5. *Osmotic Pressure*

It was observed a long time ago that when a cylinder containing wine was covered with animal membranes and placed in a bucket of water, the bladder swelled. The increased pressure in the tube is called osmotic pressure.

The following is a more common approach to study osmosis. Figure 6.5 shows a container divided into two compartments. One compartment contains a solution of solute and solvent, the other compartment contains the pure solvent. The compartments are separated by a semi-permeable membrane, which allows solvent molecules to move freely between the compartments but forbids the solute molecules to move across the boundary. Solvent molecules will diffuse into the solution causing the liquid in that compartment to rise. If pistons are placed on both compartments, the pressure on the pure solvent compartment is the atmospheric pressure P_A.

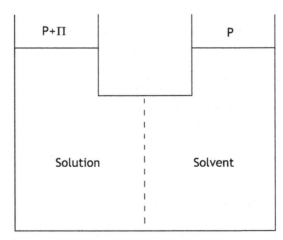

Fig. 6.5 Schematic diagram of osmotic pressure determination Π is the extra pressure that must be applied to prevent the solvent from flowing into the solution compartment.

The piston on the solution compartment will rise. To prevent that piston to rise, extra pressure, say Π, must be applied. *Osmotic pressure is the extra pressure that must be applied to the solution to prevent the flow of solvent.*

It turns out that, to good approximation, $\Pi = [B]RT$, where $[B] = n_B/V$.

Proof. At equilibrium, the chemical potential of the pure solvent is μ_A^*, and of the solvent component A in solution is μ_A. These two chemical potentials differ not only because the concentrations differ, but also because their pressures are different. For the pure solvent, $x_A = 1$ and the pressure is P. For the solution the mole fraction is x_A and the pressure is $P + \Pi$.

More explicitly, what is required is

$$\mu_A\,(P + \Pi; x_A) = \mu_A^*(P; x_A = 1) \tag{6.15}$$

First consider the non-starred μ. Assuming the solution to be sufficiently dilute, so that Raoult's Law is applicable, we have

$$\mu_A\,(P + \Pi; x_A) = \mu_A^*\,(P + \Pi) + RT \ln x_A \tag{6.16}$$

Next consider the variation of the chemical potential with pressure: $d\mu = V_m dP$, where V_m is the molar volume. This is similar to the variation of G with P, namely, $dG = VdP$. Assuming that V^* is constant, we get

$$\mu_A^*\,(P + \Pi) - \mu_A^*(P) = V^* \left(\int_P^{P+\Pi} dP \right) \tag{6.17a}$$

$$= V^*\,(P + \Pi - P) = \Pi V^* \tag{6.17b}$$

Finally, since μ_A (solution) $= \mu_A$ (pure solvent), so

$$\mu_A\,(P + \Pi, x_A) = \mu_A^*$$

$$\mu_A^*\,(P + \Pi) + RT \ln x_A = \mu_A^*(P)$$

$$\mu_A^*(P) + \Pi V^* + RT \ln x_A = \mu_A^*(P) \tag{6.18}$$

and

$$\Pi = -RT/V \ln x_A \approx -RT/[V \ln(1 - x_B)]$$
$$= x_B RT/V \approx n_B/[(n_A + n_B)(RT/V_m)]$$
$$= (n_B/V)RT \approx [B]RT \tag{6.19}$$

□

6.2.6. *Chemical Reaction Equilibria*

The relation between K and $\Delta G^{\underline{o}}$ was already discussed in previous chapters. Here, we will make a few remarks which were not sufficiently emphasized before.

a) From the free energy or chemical potentials, we can derive for the reaction

$$aA + bB \cdots = cC + dD \cdots \tag{6.20}$$

producing the quilibrium constant

$$K = a_C^c a_D^d \dots / a_A^a a_B^b \tag{6.21}$$

where the activities are related to partial pressures or to concentrations, etc. The expression of K in terms of the activities is exact, but the relation of the activities to pressure, to concentration, as shown in Eqs. (5.22a–c), are approximate. Activities are equal to the mole fraction times activity coefficients, which are not easily calculated.

b) Consider the heterogeneous reaction,

$$CaCO_3(s) = CaO(s) + CO_2(g) \tag{6.22}$$

It is found that K depends only on the partial pressure $P(CO_2)$ of the gas, i.e. $K = P(CO_2)$ and not on the other substances. Reason: the activity of each solid is 1, and the activity of the gas is to a good approximation equal to the partial pressure.

c) Also, consider the equilibrium constant for the reaction

$$2H_2O(l) = H_3O^+(aq) + OH^-(aq) \tag{6.23}$$

A neutral system in equilibrium has

$$[H_3O^+] = [OH^-] = 10^{-7}\,mol/L \tag{6.24}$$

The activity of the pure liquid is essentially constant, yielding an equilibrium constant, denoted as K_w, equal to

$$K_w = 10^{-14} \tag{6.25}$$

Example 6.4. The standard enthalpies of formation of $C_6H_5COOH(s)$, $CO_2(g)$ and $H_2O(l)$ are respectively $-386.0\,kJ$, $-285.8\,kJ$ and $-393.5\,kJ$. Calculate

(a) the standard enthalpy, $\Delta H_{comb}^{\ominus}$, and
(b) the standard energy, $\Delta U_{comb}^{\ominus}$,

of combustion of $[C_6H_5COOH(s)]$.

The combustion reaction is

$$C_6H_5COOH(s) + \frac{15}{2}O_2(g) \rightarrow 7CO_2(g) + 3H_2O(l)$$

and so

(a) $\Delta H_{comb}^{\ominus} = 7\Delta H_f^{\ominus}[CO_2(g)] + 3\Delta H_f^{\ominus}[H_2O(l)] - \Delta H_f^{\ominus}[C_6H_5COOH(s)]$
$\quad = 7(-285.8) + 3(-393.5) - (-386.0) = -2795.1\,kJ$

(b) $\Delta U_{comb}^{\ominus} = \Delta H_{comb}^{\ominus} - \Delta(PV)$
$\quad \Delta(PV) = 7PV_m[CO_2(g)] + 3PV_m[H_2O(l)] - PV_m[C_6H_5COOH(s)]$
$\quad\quad - \frac{15}{2}PV_m[O_2(g)]$

V_m are the molar volumes. Neglecting the molar volumes of the solid and liquid and replacing PV by RT (assuming ideal gas behavior), we get $\Delta[PV(g)] = (7 - \frac{15}{2})RT = -\frac{1}{2}RT$.

Thus,

$$\Delta U_{comb}^{\underline{o}} = -2795.1\,\text{kJ} + \frac{1}{2}RT$$

$$= -2795.1\,\text{kJ} + \frac{1}{2} \times 8.3145\,\text{J}\,\text{K}^{-1} \times 298\,\text{K}$$

$$= [-2795.1 + 1.24]\,\text{kJ} = -2793.9\,\text{kJ}$$

Example 6.5. The standard enthalpy change of formation in the reaction

$$H_2(g) + I_2(s) \rightarrow 2HI(g) \quad \text{is } \Delta H_f^{\underline{o}} = 52.96\,\text{kJ}.$$

The standard entropy change is $\Delta S_f^{\underline{o}} = 166.36\,\text{J}\,\text{K}^{-1}$.

(a) Calculate the equilibrium constant K at $T = 298\,\text{K}$.

$$\Delta G^{\underline{o}} = \Delta^{\underline{o}}H - T\Delta S^{\underline{o}} = [52.96 - 298 \times 0.166.36]\,\text{kJ} = 3.38\,\text{kJ}$$

$$K = \exp\left(-\Delta G^{\underline{o}}/RT\right) = \exp\left[-3380\,\text{J}/(8.3145\,\text{J}\text{K}^{-1} \times 298\,\text{K})\right]$$

$$= e^{-1.366} = 0.255$$

(b) If $T' < 298\,\text{K}$, will the equilibrium constant, K', be greater or smaller than K?

$$\ln K'(T') = \ln K(T) + [\Delta H^{\underline{o}}(T)/R] \times (1/T - 1/T')$$

$\Delta^{\underline{o}}H > 0$; If $T' < T$ the quantity $(1/T - 1/T')$ will be negative. Thus, the second term in the formula will be negative and $\ln K' < \ln K$. Thus, $K' < K$.

6.2.7. *Elements of Electrochemistry. Electrochemical Cells*

Much of electrochemistry is covered in other courses, and we will only touch on the highlights which are relevant to thermodynamics.

There are two types:

(a) *Electrolytic Cells*, which produce non-spontaneous reactions by an external electrical current. Example: the production of H_2 and O_2 from H_2O.

(b) *Galvanic Cells*, which extract energy as non-PV work in spontaneous reactions. This provides the link with thermodynamics. Examples are dry battery cells, lithium-ion cells, fuel cells, etc.

6.2.8. *Half-Reactions. Redox Reactions*

A *half-reaction* is a process in which there is a transfer of electrons from one substance to another. If there is a *loss of electrons*, the half-reaction is referred to as *oxidation*; *if* there is a gain of *electrons*, the half-reaction is referred to as *reduction*. Any electrochemical reaction can be expressed as the sum of two half-reactions.

As an example, consider the redox (oxidation–reduction) reaction

$$Cu^{2+}(aq) + Zn(s) \rightarrow Cu(s) + Zn^{2+}(aq) \qquad (6.26)$$

which is assumed to be composed of the two half-reactions

$$Cu^{2+}(aq) + 2e^- \rightarrow Cu(s) \quad \text{[reduction of } Cu^{2+}\text{]} \qquad (6.27a)$$

$$Zn(s) \rightarrow Zn^{2+}(aq) + 2e^- \quad \text{[oxidation of Zn]} \qquad (6.27b)$$

If these are used in a cell (for example, the Daniell Cell) which may be characterized as

$$Zn(s)|ZnSO_4(a)\|CuSO_4(a)|Cu(s) \qquad (6.28)$$

the oxidation half-reaction will deposit electrons on the cathode pole via the external circuit.

The half-cell potentials of the reduction and oxidation reactions referred to above are respectively +0.34 and +0.76 volt, giving an overall sum value of $E^{\underline{o}} = 1.18$ volt.

If this reaction is carried out *reversibly by letting the produced current act against a maximum resistance*, the electrical work produced by the cell will be

$$w_{\text{other}} = \Delta G_{T,P} = -n\Im E^{\underline{o}}, \qquad (6.29)$$

where n is the number of moles of electrons, $\Im$ is the Faraday constant equal to 96,485 coulomb/mol^{-1} of electrons and $E^{\underline{o}}$ is the total potential of the half-cells. The electrical work, for the Cu–Zn cell is (observing that

$1\,\mathrm{J} = 1\,\mathrm{C} \times 1\,\mathrm{V})$

$$w_{\text{other}} = \Delta G_{T,P} = -n\Im E^{\underline{o}}$$

$$= -2 \times 96{,}485\,\text{coulomb} \times 1.18\,\text{volt}$$

$$= -227{,}046.60\,\mathrm{J} \approx -227\,\mathrm{kJ} \tag{6.30}$$

The foregoing illustrates that the Gibbs free energy, a state function, can be obtained from non-PV work (which is not a state function), analogous to the determination of the enthalpy $\Delta H_{P,T}$ from PV work.

Note: Although the potential of a single electrode cannot be measured, one electrode can be assigned the value 0 and the other electrode can then be assigned a value obtained from the measured total potential. The electrode chosen to be assigned zero is the hydrogen electrode.

In the above example, we used the published electrode potentials to evaluate w_{other}. Such potentials were originally obtained by measuring voltage changes under maximum resistance between the electrodes. This is done to obtain results under quasi-static (reversible) conditions.

6.2.9. *Cells at Equilibrium*

It should be emphasized that in the above example, the current was produced by the Cu–Zn reaction, but the reaction was not in a *state of equilibrium*. If the reaction is in equilibrium, $\Delta G = 0$ and therefore, w_{other} must also be zero. Thus, there would be no current flow.

Note: One can rationalize the above by recalling that, in general, $\Delta G_{T,P} \leq w_{\text{other}}$. Thus, if the reaction is to move in a forward direction, $\Delta G_{T,P}$ must be less than w_{other} and if it is in a reverse direction, $\Delta G_{T,P}$ must be greater than w_{other}. When the system is in equilibrium, w_{other} must be zero.

Finally, since $\Delta G = \Delta G^{\underline{o}} + RT \ln K$ and thus, equal to zero (why?), it follows that $0 = -n\Im E^{\underline{o}} + RT \ln K$, and so

$$\ln K = -n\Im E^{\underline{o}}/RT \tag{6.31}$$

Example 6.6. The reduction potentials for the following reactions are

$$Cd^{+2} + 2e^- \to Cd \qquad E^{\underline{o}} = -0.40\,V$$

$$Cd + 2OH^- \to Cd(OH)_2 + 2e^- \qquad E^{\underline{o}} = -0.81\,V$$

Calculate

(a) The standard cell potential at 298 K for the reaction

$$Cd^{+2} + 2OH^- \to Cd(OH)_2;$$

(b) The equilibrium constant K at 298 K for the reaction in Part (a);
(c) The electrical work produced at constant T and P when the reaction is in equilibrium.

Solutions

(a) $Cd^{+2} + 2e^- \to Cd$ $\qquad\qquad E^{\underline{o}} = -0.40\,V$

$Cd + 2OH^- \to Cd(OH)_2 + 2e^-$ $\quad E^{\underline{o}} = +0.81\,V$

$Cd^{2+} + 2OH^- \to Cd(OH)_2$ $\qquad E^{\underline{o}}_{cell} = 0.41\,V$

(b) $\Delta G^{\underline{o}} = -\nu \Im E^{\underline{o}}_{cell}; \nu = 2$ (mol of electrons)

$\Im = 96.485 \times 10^3$ coulomb mol^{-1} (C mol^{-1})

$E^{\underline{o}}_{cell} = $ potential difference in volts $(1\,V = 1\,JC^{-1})$

$\Delta G^{\underline{o}} = -2 \times 96.485 \times 10^3\,C\,mol^{-1} \times 0.41\,JC^{-1} = -79.12\,kJ\,mol^{-1}$

$K = \exp[79.12 \times 10^3\,Jmol^{-1}/(8.3145\,J\,K^{-1}mol^{-1} \times 298\,K)]$

$= 7.38 \times 10^{13}$

(c) In general, $\Delta G_{T,P} = \Delta G^{\underline{o}}_{T,P} + RT \ln K$ and $\Delta G_{T,P} \le w_{other}$.

At equilibrium $\ln K = \Delta G^{\underline{o}}/RT$ and $\Delta G_{T,P} = w_{other} = 0$. Thus, the electrical work (a form of w_{other}) is zero.

Chapter 7

Chemical Kinetics

Much of physical chemistry is concerned with chemical reactions. Thermodynamics enables us to determine whether reactions will proceed or will not proceed. Thermodynamics will also determine the conditions that must be satisfied to obtain equilibrium, but thermodynamics will not predict how fast equilibrium will be reached. Chemical kinetics does this.

Several experimental methods are available for determining reaction rates. They basically fall into two categories:

(1) Removal of successive samples from the reaction mixture — and analyzing the mixture (hard to come by if the reaction is fast).
(2) Analyzing the reaction (using physical methods such as colorimetry, spectral absorption etc.) while the reaction is proceeding. This method is continuous.

7.1. The Rates of Reactions

Reaction rates were already studied quantitatively as far back as the 1850s. For example, the rate of the reaction

$$H_2O + C_{12}H_{22}O_{11} \rightarrow C_6H_{12}O_6 + C_6H_{12}O_6$$
$$\text{sucrose} \qquad \text{glucose} \qquad \text{fructose} \qquad (7.1)$$

was found to satisfy the relation

$$d[c]/dt = k[c] \qquad (7.2)$$

where $[c]$ is the concentration of the unreacted sucrose, and k is a constant, called the rate constant.

Reaction rates are usually expressed as the rate of decrease of a given reactant. Occasionally, the rate is expressed as the rate of increase of a particular compound. For example, the rate r for the reaction

$$a\text{A} + b\text{B} \rightarrow c\text{C} + d\text{D} \tag{7.3}$$

can be written as a rate decrease

$$r = -(1/a)\text{d}[A]/\text{d}t \tag{7.4a}$$

$$r = -(1/b)\text{d}[B]/\text{d}t \tag{7.4b}$$

or, as a rate increase

$$r = +(1/c)\text{d}[C]/\text{d}t \tag{7.4c}$$

$$r = +(1/d)\text{d}[D]/\text{d}t \tag{7.4d}$$

Note that there is a difference between the rate of *reaction*, r, and the rate of *appearance* or *disappearance* of a substance. For example, the rate of disappearance of A is defined as $-\text{d}[A]/\text{d}t$; the rate of appearance of D, is defined as $+\text{d}[D]/\text{d}t$, etc. Thus, for the reaction

$$\text{N}_2(\text{g}) + 3\text{H}_2(\text{g}) \rightarrow 2\text{NH}_3(\text{g}) \tag{7.5}$$

the following applies

$$r = -\text{d}[N_2]/\text{d}t = -1/3\text{d}[H_2]/\text{d}t = 1/2\text{d}[NH_3]/\text{d}t \tag{7.6a}$$

It should also be noted that the rate of *formation* of NH_3, on the other hand, is $2r$, whereas the rates of disappearance of N_2 and H_2 are respectively, r and $3r$. Thus, the rate of appearance of NH_3 is twice the rate of disappearance of N_2 and 2/3 times the rate of disappearance of H_2.

$$\text{d}[NH_3]/\text{d}t = 2\text{d}[N_2]/\text{d}t = \frac{2}{3}\text{d}[H_2]/\text{d}t \tag{7.6b}$$

7.2. Order of Reaction

The reaction rate is often found to be proportional to the concentration of the reactants raised to some power. For example,

$$r = -(1/a)\text{d}[A]/\text{d}t = k[A]^m[B]^n \ldots \tag{7.7}$$

where k is the reaction rate constant. The exponents $m, n, \ldots$ are not necessarily the coefficients a, b, etc. that appear in the reaction, but are determined experimentally.

The order of reaction is given by the exponents of the concentrations in the rate law. Thus, in the above example, the order of the reaction with respect to A is m, with respect to B is n, etc. The *overall* order of reaction is the sum of all the exponents. When a reaction order is mentioned, it generally refers to the overall order unless otherwise stated.

7.3. Units of the Reaction Rate Constant, k

Since the rate $r = \mathrm{d}[c]/\mathrm{d}t$ has the dimention of concentration divided by time, it follows that the product $k[A]^m[B]^n \ldots$ must have the same dimention. Thus, in a first-order reaction, $[c]t^{-1} \leftrightarrow k[c]$ and so k has the dimension of t^{-1}. In a second-order reaction, $[c]t^{-1} \leftrightarrow k[c]^2$ and thus k has the dimension $[c]^{-1}t^{-1}$.

Relation of order of reaction to stoichiometric coefficients

Consider the two reactions:

$$1) \quad 2N_2O_5 \rightarrow 4NO_2 + O_2 \tag{7.8a}$$

$$2) \quad 2NO_2 \rightarrow 2NO + O_2 \tag{7.8b}$$

The rates of these two reactions are respectively

$$1) \quad r = -\frac{1}{2}\mathrm{d}[N_2O_5]/\mathrm{d}t = k[N_2O_5] \tag{7.9a}$$

$$2) \quad r = -\frac{1}{2}\mathrm{d}[NO_2]/\mathrm{d}t = k[NO_2]^2 \tag{7.9b}$$

Obviously, reaction (1) is first order; reaction (2) is second order. This clearly shows that there is no connection between the order of reaction and the stoichiometric coefficients in these reactions.

Sometimes the order of a reaction can be zero or fractional. For example, the rate law for

$$CH_3CHO \rightarrow CH_4 + CO \tag{7.10a}$$

is

$$r = k[CH_3CHO]^{3/2} \tag{7.10b}$$

In a zero-order reaction, the rate is independent of the concentration.

If a rate law is not of the form $[A]^m[B]^n[C]^o\ldots$, the reaction has no order. For example, the rate of formation of HBr (in a particular reaction to be developed later) has the form

$$\mathrm{d}[HBr]\mathrm{d}t = \{k[H_2][Br_2]^{3/2}\}/\{[Br_2] + [HBr]\} \qquad (7.11)$$

This reaction has no order.

7.4. Determination of the Rate Law

7.4.1. *Isolation Method*

This method is used when all reactants except one are in great excess so that their concentrations virtually do not change.

Example 7.1. Consider a reaction which has the rate expression $r = k[A][B]^2$. Assuming that $[B]$ is in large excess so that it can be considered constant and equal to its initial value. So, we could express the rate as

$$r = k'[A] \qquad (7.12a)$$

where

$$k' = k[B]^2 \qquad (7.12b)$$

The rate law is now effectively first order, and is called *pseudo-first order*. Had $[A]$ rather than $[B]$ been in excess, the rate would have been *pseudo-second order*.

7.4.2. *Initial Rate Method*

In this approach, one measures instantaneous rates at the beginning of several different concentrations of a particular reactant, holding all other reactants constant.

Example 7.2. Consider the rate expression $r = k[A]^m$, of unknown order m. Denoting the initial rate of a particular sample as r_1 and of another sample as r_2, and the corresponding initial concentrations as $[A]_1$ and $[A]_2$, we can write

$$\ln(r_2/r_1) = m\ln\{[A]_2/[A]_1\} \qquad (7.13)$$

from which we can obtain the order m and from r the value of k.

7.5. Integrated Rate Law

Rate laws are differential equations which must be integrated to obtain concentrations as a function of time. The integrated rate laws are directly related to experimental observables of concentration and time.

7.5.1. *First-Order Reaction*

Consider the first-order reaction

$$r = -(1/a)\,\mathrm{d}[A]/\mathrm{d}t = k[A]^1 \qquad (7.14)$$

The rate of consumption of A is

$$-\mathrm{d}[A]/\mathrm{d}t = a\,k[A] = k_{\mathrm{A}}[A] \qquad (7.15)$$

Note that it is $k_{\mathrm{A}} = ak$ and not k that enters the expression for the integrated rate,

$$-d[A]/[A] = k_{\mathrm{A}}\mathrm{d}t \qquad (7.16)$$

Integration between $t = 0$, when the concentration is $[A]_0$ and the time t, when the value is $[A]$, gives

$$\int_{A_0}^{A} \mathrm{d}[A]/[A] = -\int_0^t k_{\mathrm{A}}\mathrm{d}t' \qquad (7.17)$$

Note that $\mathrm{d}x/x = \mathrm{d}\ln x$ and so Eq. (7.17) becomes

$$\ln [A]/[A]_0 = -k_{\mathrm{A}}t \qquad (7.18a)$$

or

$$\ln [A]_0/[A] = k_{\mathrm{A}}t \qquad (7.18b)$$

$$[A] = [A]_0 e^{-kt} \qquad (7.18c)$$

Comment: An important class of first-order reactions are nuclear reactions.

In a *first-order* reaction, a plot of $\ln [A]$ vs. t gives a straight line (Fig. 7.1), whose slope is $-k_{\mathrm{A}}$. Conversely, if the order of a reaction is not known, and a plot of $\ln[A]$ vs. t gives a straight line, this indicates that the reaction is first order.

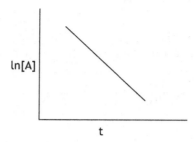

Fig. 7.1 Plot of $\ln[A]$ vs. t in a first-order reaction.

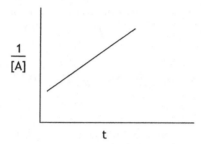

Fig. 7.2 Plot of $1/[A]$ vs. t in a second-order reaction.

In a *second-order* reaction, such as

$$-\mathrm{d}[A]/\mathrm{d}t = k_\mathrm{A}[A]^2 \qquad (7.19\mathrm{a})$$

$$-\mathrm{d}[A]/[A]^2 = k_\mathrm{A}\mathrm{d}t \qquad (7.19\mathrm{b})$$

we have $\int \mathrm{d}x/x^2 = -1/x$ and so integration of Eq. (7.19b) between the limits $[A]_0$ and $[A]$, gives

$$1/[A] - 1/[A]_0 = k_\mathrm{A}t \qquad (7.19\mathrm{c})$$

Accordingly, a plot of $1/[A]$ vs. t gives a straight line (Fig. 7.2).

Note: The foregoing does not apply to all second-order reactions. For example, for a reaction of the type $A + B \rightarrow C$ in which the concentrations of A and B are not equal, the rate law is much more complicated.

7.6. Half-Lives

The half-life of a reaction, $t_{1/2}$, is the time taken for the concentration of the substance to reduce to half its value. In a first-order reaction, from Eq. (7.18b),

$$k_A t_{1/2} = \ln\{[A_0]/[A]\} = \ln\{[A_0]/\{(1/2)[A_0]\}\} \qquad (7.20a)$$

and thus,

$$t_{1/2} = \ln 2/k_A \qquad (7.20b)$$

Example 7.3. The half-life of an enzyme-catalyzed reaction is $t_{1/2} = 138\,\text{s}$. The reaction is first order. If the initial concentration $[A]_0 = 1.28\,\text{mmol L}^{-1}$. How long will it take for the concentration to fall to $0.040\,\text{mmol L}^{-1}$.

Solution

$$k_A = \ln 2/138\,\text{s} = 5.022 \times 10^{-3}\,\text{s}^{-1}$$

Since

$$k_A t = \ln 1.28\,\text{mmol L}^{-1}/0.040\,\text{mmol L}^{-1}$$

$$= 3.4687$$

we obtain

$$t = 690\,\text{s}.$$

Note: In a first-order reaction, the half-life is independent of concentration. But in a second-order reaction, the half-life depends on the initial concentration. Exercise: Show that $t_{1/2} = 1/\{k_A[A]_0\}$.

Example 7.4. The reaction rate for the reaction

$$2N_2O_5 \rightarrow 4NO_2 + O_2$$

is of the form $r = -1/2 d[N_2O_5]/dt = k[N_2O_5]$

a) Write an expression in terms of $k[N_2O_5]$ for the rate of production of NO_2, i.e. $d[NO_2]/dt$.
b) If $k = 1.73 \times 10^{-5}\,\text{s}^{-1}$ what is the half-life of N_2O_5?

c) After a time period, t, the concentration of N_2O_5 is 10% of the initial value. What is the value of t?

Solutions

a) $r = \frac{1}{4}d[NO_2]/dt = k[N_2O_5]$
   The rate of formation of N_2O_5 is $d[NO_2]/dt = 4\,k\,[N_2O_5]$

b) The rate, r, can be written also as $r = -1/2d[N_2O_5]/dt$, and so
   $-d[N_2O_5]/dt = 2k[N_2O_5] = k_A[N_2O_5]$
   showing that the rate of reaction is of first order. Consequently,
   $t_{1/2} = \ln 2/k_A$. Thus,
   $t_{1/2} = \ln 2/(2 \times 1.73 \times 10^{-5}\,s^{-1}) = 20.0\,s$

c) $\ln[A]_0/[A] = k_A t$
   $\ln[A]_0/(0.1)[A]_0 = \ln 10 = (2 \times 1.73 \times 10^{-5})t$
   $t = 6.65 \times 10^4\,s$

7.7. Other Reaction Orders

7.7.1. *Zero-Order Reactions*

The rate decomposition of a zero-order reaction is independent of concentration,

$$-d[A]dt = k_A. \qquad (7.21a)$$

Integration gives

$$[A]_0 - [A] = k_A t \qquad (7.21b)$$

and the half-life

$$t_{1/2} = 1/2[A]_0/k_A. \qquad (7.21c)$$

A plot of $[A]$ vs. t yields a straight line with a negative slope (Fig. 7.3).

7.7.2. *Third-Order Reactions*

The rate of decomposition of a third-order reaction is

$$-d[A]/dt = k_A[A]^3 \qquad (7.22a)$$

yielding

$$-d[A]/[A]^3 = k_A t. \qquad (7.22b)$$

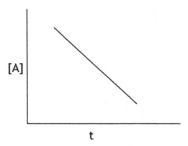

Fig. 7.3   Plot of $[A]$ vs. t in a zero-order reaction.

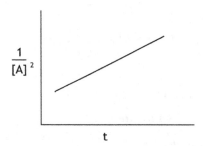

Fig. 7.4   Test for a third-order reaction.

or, integrated

$$k_{\mathrm{A}}t = \frac{1}{2}[A]^{-2} - \frac{1}{2}[A_0]^{-2} \qquad (7.22\mathrm{c})$$

Plotting $1/[A]^2$ vs. t gives a straight line of slope k_{A} (Fig. 7.4).

Exercise: Show that $t_{1/2} = 1/k_{\mathrm{A}}\{3/(2[A_0]^2)\}$.

7.8. Concentration of Products

So far, we have considered only reactant concentrations. Occasionally, product concentrations are of greater interest. In general, they are easy to obtain from the reactant concentrations.

Example 7.5. Consider the reaction $A \to P$. The P created must equal the A destroyed, i.e.,

$$[P] = [A]_0 - [A] \qquad (7.23)$$

(1) In zero-order, from Eq. (7.21b)

$$[P] = [A_0] - [A] = k_A t \qquad (7.24a)$$

(2) In first-order, from Eq. (7.18c)

$$[P] = [A_0][1 - \exp(-k_A t)] \qquad (7.24b)$$

(3) In second-order, from Eq. (7.19c)

$$1/[A] - 1/[A_0] = k_A t$$
$$[P] = k_A t [A_0]^2 / \{1 + k_A t [A_0]\} \qquad (7.24c)$$

7.9. Temperature-Dependent Reaction Rates. The Arrhenius Equation

Arrhenius found that many reactions obey the Law

$$\ln k = \ln \tilde{A} - E_a / RT$$
$$k = \tilde{A} e^{-E_a / RT} \qquad (7.25)$$

where $\tilde{A}$ is the so-called *pre-exponential* factor and E_a is the activation energy. A plot of $\ln k$ vs. $1/T$ gives a straight line whose slope is $-E_a/R$ (Fig. 7.5). If A and E_a are known, at a given T, it is easy to determine the constant at another T'. Assuming that E_a and A are constant,

$$\ln(k'/k) = (E_a/R) \; T \; (1/T - 1/T') \qquad (7.26)$$

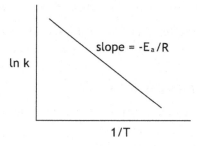

Fig. 7.5 Plot of $\ln k$ vs. $1/T$. Test of the The Arrhenius Equation.

Example 7.6. The decomposition of urea (in $0.1\,M$ HCl)

$$NH_2CONH_2 + 2H_2O \rightarrow 2NH_4^+ + CO_3^{2-}$$

is first order with rate constants at two different temperatures

Exp	$T\,(^\circ C)$	$k\,(\text{min}^{-1})$
1	61.05	0.713×10^{-5}
2	71.25	2.77×10^{-5}

Calculate A and E_a.

Solution
[Note: $\ln = 2.303 \log$]

Exp	$\log k$	$1/T$
1	-5.147	2.992×10^{-3}
2	-4.558	2.904×10^{-3}

$$E_a = \frac{2.303 \times (8.3145\,\text{J K}^{-1}\,\text{mol}^{-1})(5.147 - 4.558)}{(2.904 - 2.992) \times 10^{-3}\,\text{K}^{-1}}$$

$$= 128{,}100\,\text{J mol}^{-1}$$

$$\log A = -4.558 + 128{,}100\,\text{J mol}^{-1}/(2.303 \times 8.3145 \times 344.4\,\text{J mol}^{-1})$$

$$= 14.86678$$

$$A = 1.722$$

Arrhenius' qualitative explanation of activation energy was based on the assumption that in every system there is an equilibrium between *normal* and *active* molecules and only *active* molecules can take part in chemical reactions. According to this theory, increasing the temperature results in

(1) an increase in the number of collisions, and
(2) an increase in the concentration of active molecules.

7.10. Reaction Rate Theories

There are two theories that purport to explain reaction rates quantitatively:

(1) Collision Theory,
(2) Activated Complex Theory.

7.10.1. *Collision Theory*

This is the older of the two theories. The graph (called the reaction profile; see Fig. 7.6) depicts the potential energy of two molecules as they interact with each other. At left is the potential energy of the reactants when they approach each other. As the molecules get closer, the potential energy increases because the bond bends and starts to break. The potential energy reaches a peak when the molecules are most distorted. Thereafter the energy decreases as new bonds are formed. For reactions to succeed, molecules must collide with sufficient energy to carry them over the activation barrier.

Consider the reaction $A + B \rightarrow \cdots$. The reaction can obviously not occur more often than the number of collisions between A and B. It can be shown that the number of collisions per unit time is proportional to the product of the concentrations of A and B, i.e. rate $= k[A][B]$. However, not all collisions lead to reactions. Only the collisions which give rise to a kinetic energy exceeding E_a are effective. It can be shown (by statistical mechanics) that the fraction of molecules having a kinetic energy exceeding E_a is $e^{-E_a/RT}$. Hence, the theoretical reaction rate has to be proportional to $[A][B]e^{-E_a/RT}$. Considering that the actual reaction rate is $k[A][B]$, it follows that $k = \tilde{A}e^{-E_a/RT}$, where $\tilde{A}$ is a proportionality constant.

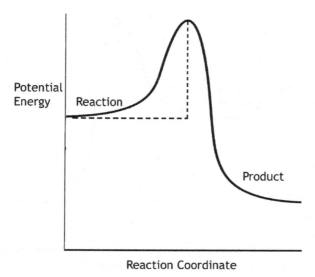

Fig. 7.6 Reaction profile of two approaching molecules.

The value of $\tilde{A}$ can, in principle, be measured. The measured $\tilde{A}$ is generally smaller than the one calculated. It is said that the relative orientation must also be taken into account, and this will give rise to a "steric" factor, P_A. Hence, $k = P_A \tilde{A} e^{-E_a/RT}$.

7.10.2. *Activated Complex Theory*

This is the modern theory of reaction rate. The collision theory is generally limited to classical gases. The activated complex theory makes use of quantum mechanical and statistical mechanical concepts.

In the activated complex theory (see Fig. 7.7), as in the collision theory, the potential energy goes through a maximum. But in the activated complex theory, the maximum corresponds to a complex molecule, which has a *definite composition* and *a loose structure*. The complex can turn into products or collapse back into reactants. An activated complex at peak potential energy is often referred to as *transition* state.

The activated complex theory was developed in 1933 by Eyring (and also by Polanyi). In essence, it assumes that there is an equilibrium involving the reactants and the activated complex, giving rise to an

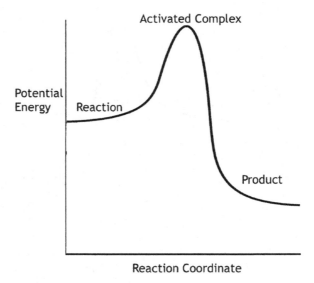

Fig. 7.7 Plot of potential energy in activated complex theory.

"equilibrium constant", defined as

$$K^{\ddagger} = [\text{activated complex}]/[\text{reactants}] \qquad (7.27)$$

Moreover, it is assumed that the rate of formation of the *product* is proportional to the concentration of the activated complex. Hence,

$$\text{Rate} = \alpha K^{\ddagger}[\text{reactants}] = \alpha K^{\ddagger}[A][B] \qquad (7.28)$$

where α is the proportionality symbol and A and B are reactants.

Comparing this rate of formation of the product with the actual formation rate $= k[A][B]$ shows that

$$k = \alpha K^{\ddagger} \rightarrow \alpha e^{-\Delta G^{\ddagger}/RT} = \alpha e^{\Delta S^{\ddagger}/RT} e^{-\Delta H^{\ddagger}/RT} \qquad (7.29)$$

This has the form of the Arrhenius equation and will be equal to it if we set

$$\Delta H^{\ddagger} = E_{a} \quad \text{and} \quad \alpha e^{\Delta S^{\ddagger}/R} = P_{a}\tilde{A}.$$

More precisely (not derived here),

$$k = (RT/Nh)e^{\Delta S^{\ddagger}/R} e^{-\Delta H^{\ddagger}/RT} \qquad (7.30)$$

in which,

$$\Delta H^{\ddagger} = E_{a} \quad \text{and} \quad P_{A}\tilde{A} = RT/Nhe^{\Delta S^{\ddagger}/R}$$

where N is the number of molecules and h is Planck's constant.

7.11. Rate Law Mechanisms

This section deals with explanations of reaction kinetics in terms of reaction mechanisms. Before discussing the reasons, it is important to distinguish between the concept of *order* and the concept of *molecularity*.

- *Order is* an empirical quantity obtained experimentally.
- *Molecularity* indicates how many molecules come together to react in an elementary reaction.

Note: In a unimolecular reaction, a single molecule breaks apart. (Radioctive decay is a good example.) The order is first order. In a bimolecular reaction, two molecules must combine but the order is not necessarily second order. It may be first order, zero order ... in fact it could be any order.

Question: Why is there no direct correspondence between molecularity and order of reaction? The reason is that most reactions are complex, not elementary. In general, reactions go through many intermediate steps.

Reactions that tend to cause complications are

$$1) \quad \text{Opposing or reverse reactions A + B} \leftrightarrow \text{C} \tag{7.31}$$

$$2) \quad \text{Complex reactions A} \begin{array}{c} \nearrow \text{B} \\ \searrow \text{C} \end{array} \tag{7.32}$$

$$3) \quad \text{Consecutive reactions A} \rightarrow \text{B} \rightarrow \text{C} \tag{7.33}$$

Such reactions greatly complicate the differential equations and, in general, exact solutions are not feasible. When there are many intermediate steps, one generally resorts to approximations.

The most common approximations are

1) Steady-State Approximation,
2) Rate-Determining (or Rate-Limiting) Step or Equilibrium Approximations.

7.12. The Steady State Approximation

Complex reactions usually involve intermediate species that do not appear in the overall reaction. If $[A]$ represent a reactant, $[P]$ a product and $[I]$ an intermediate, then it is generally true that $[I]$ is much smaller than $[A]$ or $[P]$. The reason is that intermediates are reactive, and generally do not accumulate to any extent. It is therefore assumed that $d[I]/dt = 0$. This is the steady-state assumption.

Example 7.7. Consider the reaction

$$2NO(g) + O_2(g) \rightarrow 2NO_2(g) \tag{7.34}$$

It is found experimentally that the reaction obeys the rate law

$$d[NO_2]/dt = k[NO]^2[O_2] \tag{7.35}$$

Caution. It is important to distinguish between rate of reaction and rate of consumption or rate of production. Thus, if r *represents the rate of reaction,* then

$$r = -\frac{1}{2}d[NO]/dt = -d[O_2]/dt = \frac{1}{2}d[NO_2]/dt \tag{7.36}$$

On the other hand, the rates of consumption of NO and O_2 are respectively

$$-d[NO]/dt = 2r \tag{7.37a}$$

and

$$-d[O_2]/dt = r \tag{7.37b}$$

and the rate of formation of NO_2 is

$$+d[NO_2]/dt = 2r \tag{7.37c}$$

Obviously, the rate law, r, does not represent the rate of formation of the product NO_2.

To explain the observed rate law, the following mechanisms have been suggested. It is assumed that N_2O_2 is the intermediate and that the following elementary reactions take place. [Note that reactions are said to be *elementary*, when the order of the reactions can be determined from the stoichiometric coefficients.]

1) $NO + NO \rightarrow N_2O_2$
$$d[N_2O_2]/dt = k_1[NO]^2 \tag{7.38}$$

2) $N_2O_2 \rightarrow NO + NO$ fast
$$-d[N_2O_2]/dt = k_1'[N_2O_2] \tag{7.39}$$

3) $N_2O_2 + O_2 \rightarrow 2NO_2$ slow
$$-d[N_2O_2]/dt = k_2[N_2O_2][O_2] \tag{7.40}$$

Note: These k's are not the k_A which are the product of k and a.

The overall rate for $d[N_2O_2]/dt$ is zero. So,

$$k_1[NO]^2 - k_1'[N_2O_2] - k_2[N_2O_2][O_2] = 0 \qquad (7.41)$$

Hence,

$$[N_2O_2] = k_1[NO]^2/\{k_1' + k_2[O_2]\} \qquad (7.42)$$

The product NO_2 is produced only in reaction (3), which is the slow reaction and thus the reaction determining step. For this reaction,

$$r = \frac{1}{2}d[NO_2]/dt = k_2[N_2O_2][O_2] \qquad (7.43)$$

and thus the rate of production of NO_2 is

$$d[NO_2]/dt = 2k_1k_2[NO]^2[O_2]/(k_1' + k_2[O_2]) \qquad (7.44)$$

Finally, assuming that k_1' is much greater than k_2, we can neglect the term $k_2[O_2]$ in the denominator, yielding

$$d[NO_2]/dt = 2(k_1k_2/k_1')[NO]^2[O_2] = k[NO]^2[O_2] \qquad (7.45)$$

which is the expression obtained experimentally.

7.13. The Rate-Determining Step (or Equilibrium) Approximation

Here, the reaction mechanism is assumed to consist of one or more reversible reactions that *stay close to equilibrium* during most of the reaction. This reaction is followed by a slow step which, as noted before, is the rate-determining step.

Example 7.8. Consider the reaction catalyzed by Br^-

$$H^+ + HNO_2 + C_6H_5NH_2 \xrightarrow{Br^-} C_6H_5N_2^+ + 2H_2O \qquad (7.46)$$

The observed reaction rate law is

$$r = k[H^+][HNO_2][Br^-] \qquad (7.47)$$

Proposed mechanisms:

$$(1) \ \ H^+ + HNO_2 \rightleftarrows H_2NO_2^+ \quad \text{rapid equil.} \qquad (7.48)$$

$$(2) \ \ H_2NO_2^+ + Br^- \rightarrow ONBr + H_2O \quad \text{slow} \qquad (7.49)$$

$$(3) \ \ ONBr + C_6H_5NH_2 \rightarrow C_6H_5N_2^+ + H_2O + Br^- \quad \text{fast} \qquad (7.50)$$

The rate constants for reactions (7.48)–(7.50) are respectively

(1) k_1 (forward $\rightarrow$) and k_1' (reverse $\leftarrow$),
(2) k_2 (forward $\rightarrow$) ,
(3) k_3 (forward $\rightarrow$)

The rate-determining step (Step 2) gives

$$r = k_2[H_2NO_2^+][Br^-] \qquad (7.51)$$

From (1) we get the equilibrium constant

$$K = k_1/k_1' = [HNO_2^+]/\{[H^+][HNO_2]\} \qquad (7.52)$$

which yields,

$$r = k_2K[H^+][H_2NO_2][Br^-] \qquad (7.53a)$$

or

$$r = (k_2k_1/k_1')[H^+][HNO_2][Br^-] \qquad (7.53b)$$

Equating the empirical constant k with k_2k_1/k_1' gives the observed reaction rate law.

7.14. Unimolecular Reactions

First order gas phase reactions are usually called unimolecular reactions. But for a molecule to break up, it must acquire enough energy. How does it do that? By colliding with another molecule. But that is a bimolecular reaction. The overall reaction, it turns out, has both bimolecular and unimolecular steps.

7.14.1. *The Lindemann Mechanism*

This was the first successful explanation of a unimolecular reaction.

It is assumed that a molecule A collides with another molecule A to produce an energetically excited molecule A*. The following reactions (assumed to be elementary) occur. In all these reactions, the rate constants k_1 and k_2 are in the forward direction, and k_1' is in the reverse direction.

$$1) \quad A + A \rightarrow A^* + A$$

$$d[A^*]/dt = k_1[A]^2 \tag{7.54}$$

The A* molecule may lose its energy by colliding with another molecule A.

$$2) \quad A^* + A \rightarrow A + A$$

$$-d[A^*]/dt = k_1'[A^*][A] \quad \text{fast} \tag{7.55}$$

or A* may shake itself apart, by undergoing a unimolecular reaction

$$3) \quad A^* \rightarrow P$$

$$-d[A^*]/dt = k_2[A^*] \quad \text{slow} \tag{7.56}$$

Assuming that A^* is the intermediate, and thus

$$d[A^*]/dt = 0$$

$$= k_1[A]^2 - k_1'[A^*][A] - k_2[A^*] \tag{7.57a}$$

which gives

$$[A^*] = k_1[A]^2 / \{k_2 + k_1'[A]\} \tag{7.57b}$$

Finally, if (not always true) reaction (3) is very much slower than (2), and thus $k_1' \gg k_2$, we obtain

$$[A^*] = k_1/k_1'[A] = k[A\} \tag{7.58a}$$

and therefore,

$$-d[A^*]/dt = d[P]/dt = k[A] \tag{7.58b}$$

which is first order.

7.15. Chain Reactions

Many liquid and gaseous reactions are chain reactions, meaning that an intermediate produced in one step generates an intermediate in a subsequent step; the latter generates another intermediate, etc.

It is customary to characterize the various reactions by names, such as *initiation* step, *propagation* step, *termination* step, etc.

As an example, consider the reaction

$$H_2(g) + Br_2(g) \rightarrow 2HBr(g) \tag{7.59}$$

The rate law is found to be

$$d[HBr]/dt = k[H_2][Br_2]^{3/2}/([Br_2] + k'[HBr]) \tag{7.60}$$

The proposed mechanism involves chain reactions and free radicals. A radical is denoted by a dot after the atomic symbol. The final results must not contain intermediate free radicals. Again, all reaction constants are in the forward direction.

1) *Initiation*

$$Br_2 \rightarrow 2Br\bullet$$

$$1/2\,d[Br\bullet]/dt = k_a[Br_2] \tag{7.61}$$

2) *Propagation*

$$Br\bullet + H_2 \rightarrow HBr + H\bullet$$

$$-d[Br\bullet]/dt = d[H\bullet]/dt = k_b[Br\bullet][H_2] \tag{7.62}$$

$$H\bullet + Br_2 \rightarrow HBr + Br\bullet$$

$$-d[H\bullet]/dt = d[Br\bullet]/dt = k_b'[H\bullet][Br_2] \tag{7.63}$$

3) *Retardation*

$$H\bullet + HBr \rightarrow H_2 + Br\bullet$$

$$-d[H\bullet]/dt = d[Br\bullet]/dt = k_c[H\bullet][HBr] \tag{7.64}$$

4) *Termination*

$$Br\bullet + Br\bullet + M \rightarrow Br_2 + M$$

$$-\frac{1}{2}d[Br\bullet]/dt = k_d[Br\bullet]^2 \tag{7.65}$$

Here, M is a third body which removes energy. The concentration of M has been absorbed in the reaction rate constant, k_d. (There are other recombination reactions, such as $2H\bullet \to H_2$ and $H\bullet + Br\bullet \to HBr$, but these are unimportant and will be ignored.)

The final reaction rate must not contain free radicals $Br\bullet$ or $H\bullet$. Using the steady state approximation, we consider

$$d[Br\bullet]/dt = 2k_a[Br_2] - k_b[Br\bullet][H_2] + k_b'[H\bullet][Br_2] + k_c[H\bullet][HBr]$$

$$- 2k_d[Br\bullet]^2 = 0 \qquad (7.66)$$

$$d[H\bullet]/dt = k_b[Br\bullet][H_2] - k_b'[H\bullet][Br_2] - k_c[H\bullet][HBr] = 0 \qquad (7.67)$$

Thus, there are two equations with two unknowns. Solving them gives

$$[Br\bullet] = (k_a/k_d)^{1/2}[Br_2]^{1/2} \qquad (7.68)$$

$$[H\bullet] = \{k_b(k_a/k_d)^{1/2}[H_2][Br_2]^{1/2}\}/\{k_b'[Br_2]k_c[HBr]\} \qquad (7.69)$$

The final expression for the rate law, which is in terms of the production of HBr, combines Eqs. (7.62)–(7.64) to give

$$d[HBr]/dt = k_b[Br\bullet][H_2] + k_b'[H\bullet][Br_2] - k_c[H\bullet][HBr] \qquad (7.70)$$

Substitution of the values for $[H\bullet]$ and $[Br\bullet]$ gives

$$d[HBr]/dt = 2k_b(k_a/k_d)^{1/2}[H_2][Br_2]^{3/2}/\{[Br_2] + (k_c/k_b')[HBr]\} \qquad (7.71)$$

which, when replacing k by $2k_b(k_a/k_d)$ and k' by k_c/k_b', gives the experimental rate law.

Example 7.9. The production of HI in the reaction

$$H_2 + I_2 \to 2HI$$

obeys the rate law $d[HI]/dt = k[H_2][I_2]$.

The following mechanism has been suggested

$$1)\ I_2 \rightleftarrows 2I \quad \text{rapid equilibrium}$$
$$2)\ 2I + H_2 \to 2HI \quad \text{slow}$$

The forward and reverse reaction constants of Eq. (1) are k_1 and k_1'; the (forward) reaction constant of Eq. (2) is k_2. Determine the rate law $d[HI]/dt$ and express the result in terms of the reaction constants k_1, k_1', k_2.

Solutions

a) Equilibrium Approximation:

$$k_1[I_2] = k_1'[I]^2; \quad K = [I]^2/[I_2] = k_1/k_1'; \quad [I]^2 = (k_1/k_1')[I_2]$$

The rate-determining step is Step (2);

$$r = \tfrac{1}{2}d[HI]dt = k_2[I]^2[H_2]$$
$$d[HI]/dt = 2k_2[I_2] = (2k_2k_1/k_1')[I_2][H_2]$$

b) Steady-State Approximation:

The intermediate substance, I, is

formed in Step (1) forward direction: $\tfrac{1}{2}\,d[I]/dt = k_1[I_2]$
reduced in Step (1) reverse direction: $-\tfrac{1}{2}\,d[I]/dt = k_1'[I]^2$
reduced in Step (2) $-\tfrac{1}{2}\,d[I]/dt = k_2[I]^2[H_2]$
Since the intermediate substance must disappear, we have

$$k_1[I_2] - k_1'[I]^2 - k_2[I]^2[H_2] = 0$$

and so $[I]^2 = k_1[I_2]/(k_1' + k_2[H_2])$. Thus, using Step 2 gives

$$d[HI]/dt = 2k_2[I_2][H_2] = 2k_2k_1[I_2][H_2]/(k_1' + k_2[H_2])$$

Realizing that Step (2) is the slow step, it means that $k_2 \gg k_1'$ and so neglecting the k_2 term in the denominator gives

$$d[HI]/dt \approx (2k_2k_1/k_1')[I_2][H_2]$$

which is the same as in Part (a).

Chapter 8

Introduction to Quantum Theory

This chapter traces the historical development that led to the adoption of quantum theory. In subsequent chapters, quantum theory will be applied to atoms and molecules, to translational, rotational and vibrational motion, and to spectroscopy.

Specifically, the present chapter deals with the following topics:

(1) Failure of classical mechanics
(2) Wave-particle duality
(3) The Schrödinger equation and the Born interpretation of wave-functions
(4) The Uncertainty Principle and the Superposition of State
(5) Structure of atoms — atomic orbitals, energy levels, atomic spectra, *aufbau* (build-up) *principle*, radius and ionization energy

8.1. Historical Development

Around 1900, science in general and physics in particular were considered to be a self-contained, elegant disciplines. All known phenomena could be explained, in principle, by what is now known as *classical* theory. Specifically, the subjects that were well developed were

(1) Mechanics (dealing with motion)
(2) Light (electromagnetic radiation), electricity and magnetism
(3) Thermodynamics and statistical mechanics

However, when these classical theories were applied to systems of microscopic dimension (atoms, molecules, etc.), they generally failed. To be cognizant of the confusion and disappointment these failures caused, it is well to mention some of the disputes scientists engaged in.

In the 17th century there were two theories of light:

(1) The corpuscular theory (proposed by Isaac Newton), which stated that light consists of "particles" shot out of the luminous body.
(2) The wave theory of light (proposed by Huygens), suggesting that light consists of waves.

First, the corpuscular theory was most favored as it could account for the sharp shadows, but as time went on, the appearance of diffraction patterns when light was passed through two adjacent orifices (Young's Experiment) and the prediction of Maxwell's Theory of Electromagnetic Radiation, clearly favored the wave theory.

In Young's experiment ($\sim$1850), the light emanating from the two holes was observed on a screen. If light were corpuscular, one should observe two lit points on the second screen. If light were wave-like, the light emanating from the two holes would set up new waves that would interfere with each other (constructively and destructively), thus producing interference patterns. The experiment clearly showed the latter, establishing the wave theory as the proper one.

Furthermore, Maxwell's Theory of Electromagnetic Radiation, which formulated a unified treatment of electricity, magnetism and light (called electromagnetic radiation), clearly established light to be wave-like. By the end of the 20th century, the particle theory of light was as dead as a doornail.

8.2. Failure of Classical Theories

Important experiments which defied classical explanations around 1900 were *black-body radiation, photo-electric effect* and *heat capacity of solids.*

8.2.1. *Black-Body Radiation*

When a body is heated to incandescence, light is emitted. First, dull red light is emitted. Then, as temperature is increased, bright red light is emitted. Thereafter yellow light etc. is given off and eventually blue light is emitted. The distribution of the light colors (really in terms of frequencies or wavelengths) could be measured by shining the emitted light onto a prism.

The prism separates the light, which is a mixture of several frequencies (colors), thus providing information on the distribution of the frequencies (or colors). The theoretical prediction of the way frequencies should vary with temperature did not agree with the experimental results.

Classically, it was assumed that the electromagnetic field is made up of a collection of oscillators. If an oscillator is excited to a frequency ν it will emit radiation of that frequency. Planck noticed that if he would disregard one of the most sacred rules of science, namely that *energy varies continuously*, he could explain the black-body radiation.

Planck then postulated that the oscillator *energies varied discontinuously — that the energies are discrete, not continuous*. Basically, Planck suggested that in the black-body experiment, energy can be absorbed from the heat source only *in discrete units, called quanta*.

8.2.2. *Photo-Electric Effect*

When a beam of light shines on a metal, it may cause electrons to be ejected from the metal. This is referred to as the photo-electric effect. However, there is a minimum (or threshold) frequency (different for different metals) below which electrons are not ejected, regardless of how intense the light beam is.

Note: If light consisted of waves, as was generally assumed in those days, there was no reason why there ought to be threshold frequency. As long as the beam is sufficiently intense, it should emit electrons regardless of frequency.

Could Planck's ideas shed some light on this problem? Planck's suggestion of 1900 was essentially forgotten until Einstein, a patent clerk at Bern, Switzerland, published a paper in 1905, which not only revived Planck's ideas but plunged science into a state of turmoil that lasted for more than a quarter of a century.

Einstein felt that Planck's explanation of the black-body radiation, though revolutionary, was incomplete. While Planck's theory clearly suggested that energy absorbed from the heat source by the black-body must do so in quanta, not much was said about the fate of the energy of the emitted radiation. It was generally assumed that the emitted radiation would obey the Laws of Maxwell, in other words, would be wave-like. But Einstein argued that these ideas are inconsistent. He maintained that if energy is absorbed from the heat source in quanta, it must also emit radiation energy in the form of quanta, clearly implying that radiation is particle-like.

Einstein then introduced the notion that *light consists of indivisible units* (later called photons), the energy of which is $\varepsilon = h\nu$, where h is Planck's constant and ν the frequency. Planck's constant has the value $h = 6.626 \times 10^{34}$ J s or $h = 6.626$ kg m^2 s^{-1}.

> Comment: It is ironical that Einstein who brought down Newton's classical mechanics with relativity resurrected Newton's corpuscular theory of light.

Einstein's idea was not based strictly on logic, as compelling as that was; but by postulating the existence of indivisible light units, he was able to explain the photo-electric effect.

According to Einstein, electrons in a metal are knocked out by photons. The energy of a photon must be equal to or exceed the attractive energy between the electron and the positive charge of the metal. This energy is called the *work-function* and is denoted as ϕ. If there is excess energy, it is carried off by the emitted electron in the form of kinetic energy. Thus, $KE_{\text{elecr}} = \varepsilon_{\text{phot}} - \phi$.

8.2.3. *Heat Capacity of Solids*

Solids consist of atoms which are oscillating. According to classical theory, the average energy of an oscillator in 3-D is $3kT$, where k is the Boltzmann constant equal to R/N_A, (N_A being Avogadro's number). The vibrational molar energy is $3RT$ and the heat capacity is $C_V = 3R$ (the value of Dulong and Petit). The actual heat capacity, as shown in Fig. 8.1, is constant with a value of $3R$; but as the temperature is lowered, the heat capacity decreases, and at 0 K it is zero.

To explain this, Einstein invoked Planck's hypothesis that the energy of an oscillator has to be quantized. This produced a heat capacity curve which is very similar to the observed one. The formula was further improved by Debye who did essentially what Einstein did but considered the solid to vibrate with a range of frequencies rather than a single frequency as Einstein had done.

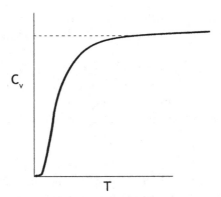

Fig. 8.1　Plot of variation of C_V with T.

8.2.4. *Wave or Particle?*

So what is light? The photo-electric effect clearly established light to be particle-like. Young's experiment on interference diffraction and Maxwell's theory of electromagnetic radiation clearly established light to be wave-like. How could the same entity, light, be such a contradictory thing?

8.3. The Rutherford Atom

The Rutherford experiment ($\sim$1911) is one of the most important experiments ever performed, for it established unequivocally the structure of the atom (Fig. 8.2).

(1) *The experimental set-up.* In this experiment, a narrow beam of α rays from a radioactive source (radium) was aimed at a very thin gold foil. Behind the gold foil was a screen coated with a chemical (zinc sulfide), which has the following property: it flashes when hit by an α-particle.

Note: Some naturally occurring heavy atoms (e.g. uranium, radium, polonium, etc.) have the ability to emit rays. The rays are of three types: α-rays, β-rays and γ-rays. The α-rays are bare positively charged helium nuclei (helium atoms stripped of the electrons); β-rays are electrons; γ-rays are the most energetic forms of electromagnetic radiation.

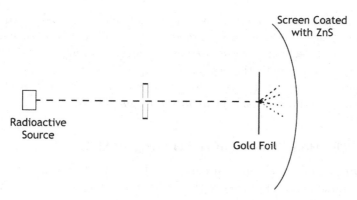

Fig. 8.2 The Rutherford experiment.

(2) *Observation.* What Rutherford observed was that most of the α-particles passed through the gold foil undeflected. (He could tell by observing the scintillations on the screen.) Some α-particles were deflected.

(3) *Interpretation.* The results led to the following conclusions:

   (a) The gold foil is for the most part empty space since most of the α-particles passed through the foil undeflected.

   (b) The few α-particles that passed through deflected indicated that most of the atomic mass is concentrated in a tiny nucleus.

   (c) The gold nucleus must be positively charged. Only positive nuclei can cause positively charged α-particles to deflect from the original path.

   (d) α-particles colliding with electrons cause no noticeable effect.

The Planetary Model of the Atom. What emerged from the Rutherford experiment was a model that in many ways resembles the solar system. In fact, the Rutherford model was referred to as the *planetary model of the atom.* In the solar system, the planets revolve around the sun. They have to be in motion; otherwise they would be pulled into the sun by the force of gravity. Similarly, in atoms, the positively charged nucleus and negatively charged electrons attract each other. Again, the electrons must be in motion; otherwise they would fall into the nucleus.

The fact that an atom behaved so similar to our solar system must have been most pleasing to scientists, but there is an essential difference between planets moving around the sun and electrons moving around a nucleus. The sun and planets are neutral (not electrically charged) bodies; the nucleus and electrons are electrically charged. And, as noted before, when a charged particle (an electron in this case) moves in an electromagnetic field, it should emit radiation (light), thereby losing energy. Thus, when an electron moves around the nucleus it should lose energy and eventually spiral into the nucleus. In other words, *a planetary atom should not exist.*

8.4. The Bohr Theory of the Hydrogen Atom

Bohr developed a theoretical model for Rutherford's atom, using Planck's idea of quantization of energy (Fig. 8.3). He further postulated that

(1) An electron in an atom is limited to certain discrete values E_1, E_2, etc. called stationary states of levels.
(2) An electron moving in a stationary state does not emit electromagnetic radiation.
(3) Radiation is emitted or absorbed only when an electron jumps from one level to another. The energy of the photon is $\varepsilon_{\text{photon}} = h\nu = (E_n - E_m)$, E_n and E_m being two energy levels of the hydrogen atom. If $E_n > E_m$, a photon is emitted; if $E_n < E_m$, a photon is absorbed.
(4) The electron moves in circular orbits (paths), and
(5) There were certain limitations of angular momentum, requiring energy levels to have the form

$$E_n = -hcR_{\text{H}}/n^2 \quad n = 1, 2, \ldots \tag{8.1}$$

where c is the speed of light and R_{H} is the Rydberg constant for the hydrogen atom, namely $R_{\text{H}} = 1.0946 \times 10^{-5}\,\text{cm}^{-1}$.

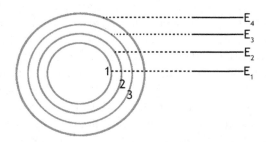

Fig. 8.3   Diagram depicting orbitals and energy levels of the Bohr atom.

With these assumptions, Bohr was able to generate the well-known empirical formulas for the emission of radiation by excited hydrogen atoms. They are

$$\nu/c = R_{\mathrm{H}} \left(1/1^2 - 1/n^2\right) \quad n = 2, 3, \ldots \quad \text{Lyman Series} \qquad (8.2a)$$

$$\nu/c = R_{\mathrm{H}} \left(1/2^2 - 1/n^2\right) \quad n = 3, 4, \ldots \quad \text{Balmer Series} \qquad (8.2b)$$

$$\nu/c = R_{\mathrm{H}} \left(1/3^2 - 1/n^2\right) \quad n = 4, 5, \ldots \quad \text{Paschen Series} \qquad (8.2c)$$

Note: Bohr's theory was essentially classical in nature except for the assumption of quantization of energy (really of angular momentum of the moving electron) and the photon nature of radiation. This forced him to postulate the conditions 4 and 5.

Bohr's prediction of radiation of the hydrogen atom was in almost perfect agreement with experiments, suggesting that the theory of he atom was solved. But when the theory was applied to He, the results were not good, and when applied to heavier atoms, the theory failed miserably.

The phenomenal success of the Bohr theory of the hydrogen atom, followed by swift decline, was so disturbing that many scientists were beginning to have serious doubts about the generality of science.

8.5. Louis de Broglie

Then, in 1923, Louis de Broglie, a physics student at the Sorbonne, submitted a dissertation in which he made a startling discovery. Rummaging through Einstein's theory of relativity, he concluded that matter must consist of waves — matter waves.

His reasoning (highly simplified here) may have been based on the following kinds of analogies: For photons, $\varepsilon = h\nu$. Photons have zero rest mass, but not when in motion. Using Einstein's relativity formula $E = mc^2$ and his expression for the photon energy $\varepsilon = h\nu = mc^2$ produced

$$h\nu/c = mc = p, \qquad (8.3a)$$

p being the momentum. Thus, for photons

$$h/mc = h/p = c/\nu = \lambda \qquad (8.3b)$$

de Broglie suggested that a similar expression applies to a particle moving with a velocity v

$$\lambda = h/mv = h/p \tag{8.3c}$$

In other words, a particle moving with a velocity v has a wave associated with it.

The de Broglie hypothesis was confirmed a few years later by Davisson and Germer, who analyzed diffraction patterns of electrons by crystals and observed interference patterns similar to inference by light.

For example, a person weighing 70 kg moves with a velocity of $v = 1\,\mathrm{m\,s}^{-1}$. By the de Broglie equation, the wavelength is

$$\lambda = 6.626 \times 10^{-34}\,\mathrm{m}^2\,\mathrm{s}^{-1}/70\,\mathrm{kg\,m\,s}^{-1} = 9.5 \times 10^{-36}\,\mathrm{m}.$$

The new ideas that were introduced next were so strange and weird that only the most gifted scientists could understand them. What finally emerged was not just a generalization of classical theories but an entirely new framework, different conceptually and philosophically from the old theories.

Three different formulations of the new theory emerged between 1925 and 1927. They are referred to as

(1) Wave-Mechanics (Schrödinger)
(2) Matrix Mechanics (Heisenberg, Born, Jordan)
(3) Quantum Theory (Dirac)

The three formulations looked mathematically different, but turned out to be equivalent. Following are some anecdotal details of the new developments.

(1) There is a story that Debye at Zurich Polytechnic said to Schrödinger (a younger faculty member), "Erwin, since you are not doing anything useful these days why don't you find out what de Broglie is up to and tell us about it." Schrödinger did that. It became clear to Schrödinger that if there was such a thing as matter waves, there must be a *wave-equation* describing those waves, analogous to the Maxwell equations describing electromagnetic waves. Schrödinger set out to discover the matter wave-equation.
(2) Heisenberg went to Helgoland trying to get rid of a severe bout of hay-fever. There he worked on problems involving quantum theory. Rather than focusing on de Broglie's ideas, as Schrödinger did, he

went back to the Bohr theory of the hydrogen atom. The phenomenal success of Bohr's theory of the hydrogen atom could not be entirely accidental, he reasoned. There must be some essential validity to that approach. In Bohr' theory, there are concepts that can be measured and concepts that cannot be measured. Heisenberg decided to develop a theory based solely on concepts that have been measured. The Bohr theory uses such quantities as energy levels, frequencies, intensities and orbits. Heisenberg argued that all we know for sure about the atom are such things as frequencies, intensities of the emitted light. No one has seen *orbits* — out they go. In fact, no one has seen energy levels, only changes in energy levels.

Heisenberg found that he could obtain the correct results for the harmonic oscillator and the hydrogen atom by replacing the classical dynamical variables, which depend on single numbers, by matrices which are arrays of numbers. He even worked out the mathematics on how to handle multiplication of arrays of numbers, not knowing that mathematicians had done that ∼100 years earlier under the name of matrix multiplication.

When Heisenberg returned to Gottingen, he showed his work to Born, who immediately recognized that Heisenberg's work was not a mere mathematical scheme for handling arrays of numbers (which was not new), but what Heisenberg actually had developed laid the foundation of a new and different theoretical framework of mechanics.

Together with Jordan, a young faculty member at Gottingen, *Heisenberg, Born and Jordan formulated a general theory of quantum mechanics which, because of the form, is also referred to as matrix mechanics.*

(3) By now Dirac came on the scene. Heisenberg, on a visit to Cambridge University, showed his work to Prof. Fowler, asking him for his opinion. Fowler gave the material to Dirac, a student of Fowler, and asked him for comments. Dirac not only did that, but formulated a quantum theory in terms of *operators.*

Dirac's formulation is *lean* in comparison with Heisenberg's, accomplishing with ease what Heisenberg, Born and Jordan could accomplish only with great difficulty. *Dirac's formulation is most elucidating in its generality and is sometimes referred to as symbolic quantum mechanics.*

8.6. The Schrödinger Equation

Schrödinger's treatment is mathematically the simplest, and is most often used in calculations. Basically, it is a differential equation which has to be solved to produce the properties of the system under consideration. For example, the Schrödinger equation of a particle moving in one dimension, having a mass m, is

$$-(h^2/8\pi^2 m)\,d^2\psi/dx^2 + V\psi = E\psi \qquad (8.4)$$

where V is the potential energy, E is the energy of the particle and ψ is the wave-function. Solving this equation produces two types of information:

(1) A set of energy levels: E_1, E_2, etc.
(2) A set of wave-functions, $\psi_n\,(x)$, associated with each energy state.

Originally, Schrödinger believed that the wave function described the position of the particle. However, Bohr, who invited him to Copenhagen, quickly convinced him that that was not the case. The wave-function ψ has no physical significance. Today's accepted interpretation (due to Born) is that $\psi^*\psi$ represents a *probability density*, and that $\psi^*\psi\delta v$ denotes the *probability of finding the particle within the volume element δv*. [The wave-function is in general complex and the ψ^* denotes the complex conjugate. The product of the wave-function and its complex conjugate is always real.]

Example 8.1. The ground state of He$^+$ is

$$\psi = (8/\pi a_0^3)^{1/2} \exp(-2r/a_0^3) \qquad (8.5)$$

where r is the distance of the electron from the nucleus, and

$$a_0 = 52.9 \times 10^{-12}\,m = (52.9\,pm) \qquad (8.6)$$

(a) What is the probability of finding the electron within a volume of $\Delta v = 1\,pm^3$ around the nucleus?
(b) What is the probability of finding the electron within a volume of $1\,pm^3$ at a distance a_0 from the nucleus?

Solution
The probability expression is (ψ is real)
Prob $= \int \psi\psi dv = \int (8/\pi a_0^3)\,\{\exp(-2r/a_0^3)\}^2\,dv$
In this problem, the r's are constant and $\int dv = 1\,pm^3$.

(a) At the nucleus, $r = 0$ and

$$\begin{aligned}
\text{Prob} &= (8/\pi a_0^3) \times (1\text{pm})^3 \\
&= \{8/(\pi \times 52.9^3 \text{ pm}^3)\}(1\text{pm}^3) \\
&= 1.72 \times 10^{-5}
\end{aligned}$$

(b) At $r = a_0$,

$$\begin{aligned}
\text{Prob} &= (8/\pi a_0^3) \times \{\exp(-2a_0/a_0)\}^2 (1\text{pm})^3 \\
&= \{8/(\pi \times 52.9^3 \text{ pm}^3)\}(\exp - 2)^2 \text{ pm}^3 \\
&= 3.15 \times 10^{-7}
\end{aligned}$$

The probabilistic nature of quantum mechanics manifests itself in

(a) The Schrödinger formulation through the wave function.
(b) The Heisenberg formulation through the *Uncertainty Principle*, which states that it is not possible to know with complete certainty both the *momentum and position of a particle*. More precisely, if Δp and Δx represent respectively the uncertainty in momentum and the uncertainty in position, then quantum theory requires that

$$\Delta p \times \Delta x \geq h/4\pi \tag{8.7}$$

8.7. Summary and Conclusions

(1) Quantum mechanics is *probabilistic*, unlike classical mechanics, *which is deterministic*.
(2) *The Copenhagen Interpretation.* In 1927, the leading scientists met at Lake Como for the purpose of arriving at some consensus of the meaning of quantum mechanics. This resulted in an interpretation, referred to as the Copenhagen Interpretation, because of the leadership of Bohr. Most scientists of those days accepted the Copenhagen Interpretation as the true meaning of quantum mechanics. Einstein never did. Another notable scientist who had problems with some of the Copenhagen Interpretation was Schrödinger.

The most important conclusions of the Copenhagen Interpretation are

(a) The dual nature of *light* and of *matter*. They can exist either as particles or as waves.

(b) The importance of *observation or measurement*. In classical mechanics, one can separate the observation of the system from the measuring device. *In quantum mechanics, the operation of observation is built into theory and the observed results are the properties of the system as perturbed by the measurement.*

(c) The laws of quantum mechanics have meaning only if measurements are done on the system.

(d) One of the consequences of the above is that if a system can exist in several different states, the state of the system will be a superposition, that is a mixture of all these states (sometimes referred to as "ghost" states). But if a measurement is made, the wave-function "collapses" to a particular state and only that state is observed. The *measurement causes the collapse.*

Note: As an example of the contrast of observations in classical and quantum mechanics, consider the measurement of the temperature of a person with a thermometer. What really takes place is heat flowing from the person to the thermometer. So the reading on the thermometer represents the temperature of the person after the heat loss. But the amount of heat is so small, that for all practical purposes it is negligble. The same considerations apply to all measurements of large bodies in classical mechanics. However, if an observation is made on a microscopic system of say, the size of an electron, the process of measurement will have profound effect on the results, and may not be ignored. *In quantum mechanics, the operation of observation is built into theory and the results are properties of the system as perturbed by the measurement.*

8.8. Schrödinger's Cat

Schrödinger had difficulties accepting this interpretation and he published a "thought experiment", which goes under the name of "Schrödinger's Cat". In this thought-experiment, a cat is confined in a lower compartment separated from an upper compartment by a glass ceiling (Fig. 8.4). The upper compartment contains a poisonous gas. In the upper compartment,

Fig. 8.4 Schrödinger's Cat.

there is a radioactive material which emits α-particles with a half-life of 1 hour. The emitted particle triggers a contraption which releases a hammer that shatters the glass. A half-life of an hour means that there is a 50–50 chance that an α-particle was emitted and a 50–50 chance that a particle was not emitted. According to the Copenhagen Interpretation, the state of the system after an hour is in a state of *decay* and a state of *non-decay coexisting simultaneously*. Only after an observation will the combined state collapse to either the *decayed state* or the *non-decayed state*. In the decayed state, the cat is obviously dead and in the non-decayed state the cat is alive. In other words, if no observation is made, the cat is simultaneously *dead and alive*. "Does that make any sense?" Schrödinger asked.

The counter-argument was that a cat has consciousness, which acts as an observation. The question was raised: "what if it is not a cat, what if it is a rat, an amoeba, a bacterium?" The Schrödinger Cat controversy was never resolved.

Finally, before taking up applications of Quantum Mechanics in the next chapters, let us examine in some more detail the topics mentioned here, namely, the de Broglie wavelength, the Uncertainty Principle, and Probability Density.

Example 8.2. An electron moves with an energy of 10 Mev. What is the de Broglie wavelength λ?

Solution

$$\lambda = h/p; \quad E = 1/2m_ev^2; \quad p = \sqrt{(2m_eE)}; \quad J = kg\,m^2\,s^{-2}$$

$$p = \sqrt{2 \times 9.11 \times 10^{-31}\,kg \times 10^7\,eV \times 1.6 \times 10^{-19}\,J/eV}$$

$$= 1.707 \times 10^{-21}\,\sqrt{(kg\,J)} = 1.707 \times 10^{-21}\,kg\,m\,s^{-1}$$

$$\lambda = (6.626 \times 10^{-34})J\,s^1/(1.707 \times 10^{-21})\,kg\,m\,s^{-2}$$

$$= (6.626 \times 10^{-34})\,kg\,m^2\,s^{-2}\,s/(1.707 \times 10^{-21})\,kg\,m\,s^{-1}$$

$$= 0.388 \times 10^{-12}\,m = 0.39\,pm\,(pm = 10^{-12}\,m)$$

Example 8.3. The uncertainty in speed of a particle weighing $1\,g$ is $\Delta v = 10^{-6}\,m\,s^{-1}$. What is the uncertainty in position?

Solution

$$\Delta x \Delta p = h/4\pi = 1.054 \times 10^{-34}\,kg\,m^2\,s^{-2}\,s$$

$$\Delta x = \frac{h/(4\pi\Delta p) = 5.267 \times 10^{-35}\,kg\,m^2\,s^{-1}}{1.0 \times 10^{-3}\,kg \times 1.0 \times 10^{-6}\,m\,s^{-1}}$$

$$= 5.3 \times 10^{-26}\,m$$

Example 8.4. The ground-state wave-function of He^+ is $\psi = (8/\pi a_0^3)^{1/2}$ $e^{-2r/a}$, where r is the distance from the nucleus, and a_0 is the Bohr radius, $a_0 = 52.9\,pm$. What is the probability of finding the electron within volume element $\delta v = 1\,pm^3$,

(a) around the nucleus;
(b) at a distance a_0 from the nucleus.

Solution

The probability is $\psi\psi\,\delta v = (8/\pi a_0^3)(e^{-2r/a})^2\,\delta v$.

$$\text{At nucleus } r = 0, \text{ Prob.} = (8/\pi a_0^3)\,\delta v$$

$$= (8/\pi)(52.9)^{-3} \times (1)^3 = 1.72 \times 10^{-5}$$

$$\text{At } r = a_0, \text{ Prob.} = (8/\pi a_0^3)(e^{-2a/a})^2\,\delta v = 1.72 \times 10^{-5} \times e^{-4}$$

$$= 3.15 \times 10^{-7}$$

Chapter 9

Applications of Quantum Theory

9.1. Translational Motion. Particle-in-a-Box

Consider a particle of mass m in a one-dimensional box of length L. Assuming that the potential energy is $V = 0$ inside box and $V = \infty$ outside the box, the solution of the Schrödinger equation yields the energy (see Fig. 9.1):

$$E_n = n^2 h^2 / 8\, m L^2 \quad n = 1, 2, \ldots \tag{9.1}$$

and wave-function

$$\psi_n = (2/L)^{1/2} \sin(n\pi x/L) \quad n = 1, 2, \ldots \tag{9.2}$$

Note that the spacing between energy levels increases with increasing n and decreases with increasing L. When L becomes very large (of macroscopic dimensions), the energy distribution becomes practically continuous.

Example 9.1. A conjugated polyene molecule is sometimes simulated by a one-dimensional-box. If $L = 2.0\,\text{nm}$ and an electron in the box is excited from state 5 to state 6, what is the transition energy?

Solution

The energy difference between level 5 and level 6 is

$$\begin{aligned}
\Delta E_{6 \leftarrow 5} &= 11\, h^2 / (8\, m L^2) \\
&= \frac{11 \times (6.6260 \times 10^{-34})^2 \,\text{J}^2\text{s}^2}{8 \times 9.10939 \times 10^{-31}\,\text{kg} \times (2.0 \times 10^{-9})^2\,\text{m}^2} \\
&= 1.6554 \times 10^{-19}\,\text{J} \approx 1.0\,\text{eV}
\end{aligned} \tag{9.3}$$

[Note: $1\,\text{J} = 1\,\text{N}\,\text{m} = 1\,\text{kg}\,\text{m}\,\text{s}^{-2}\,\text{m} = 1\,\text{kg}\,\text{m}^2\,\text{s}^{-2}$; $1\,\text{eV} = 1.607 \times 10^{-19}\,\text{J}$]

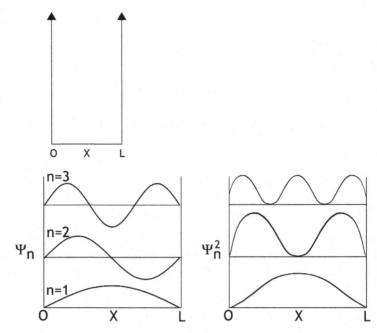

Fig. 9.1   Particle-in-a-box wave-functions.

9.2. Hydrogenic Atoms (H, He⁺, Li²⁺, etc.)

The Schrödinger equation yields the following solution:

$$E_n = -hcZ^2 R_H/n^2 \tag{9.4}$$

where Z is the atomic number and R_H is the Rydberg constant for the H atom,

$$R_H = m_e\, e^4/8\, \varepsilon_0\, h^2 \tag{9.5}$$

m_e being the mass of the electron and ε_0 the vacuum permeability. Most often, the combination Z^2 is used where R defines the *particle* Rydberg constant.

The hydrogenic wave-functions are characterized by 3 quantum numbers:

1) The *principle quantum number*, $n = 1, 2, \ldots$, which determines the energy of the electron in the atom.

2) The *azimuthal quantum number*, $l = 0, 1, 2, \ldots$, which determines the orbital angular momentum of the electron.
3) The *magnetic quantum number*, $m_l = 0, \pm 1, \pm 2, \pm 3, \ldots$, which determines the angular momentum of an electron along a particular axis.

9.3. One-Electron Wave-Functions

One-electron wave-functions of hydrogenic atoms are called *orbitals*, and are usually denoted by the symbols s, p, d, etc. representing $l = 0, 1, 2$, etc. preceded by a number which indicates the principal quantum number. When $l = 0$, the orbital is denoted as s, when $l = 1$, the orbital is denoted as p, etc. There are three different p's — p_x, p_y, p_z — which indicate the directions.

Examples:

$$\psi_{100} = (1/\sqrt{\pi})(Z/a_0)^{3/2} e^{-Zr/a_0}; \ 1s \tag{9.6}$$

$$\psi_{200} = [1/(4\sqrt{2\pi})](Z/a_0)^{3/2}(2 - Zr/a_0)e^{Zr/2a_0}; \ 2s \tag{9.7}$$

$$\psi_{210} = [1/(4\sqrt{2\pi})](Z/a_0)^{5/2}e^{-Zr/2a_0}r\cos\theta; \ 2p_z \tag{9.8}$$

9.4. Ionization Energy

The ionization energy is the maximum energy needed to remove an electron from the ground state of an atom.

Example 9.2. Knowing that the ionization for H is $I = 13.9$ eV, what is it for He^+ ?

Solution

Obviously, the only difference is Z, and so $E_{He^+}/E_H = Z^2 = 4$, yielding for $He^+ = 4 \times 13.59\,eV = 54.36\,eV$.

9.5. Shells and Subshells

(a) In hydrogen and hydrogenic atoms, the energy depends only on the principal quantum number n. In all other atoms, the energy depends also on the quantum number l. It is standard practice to refer to all electrons with the same n as belonging to the same *shell*. Often, the shells are denoted by capital letters $K, L, M, N, \ldots$, referring respectively to $n = 1, 2, 3, \ldots$

n (shell)	l (subshell)	m_l (orbital)
1 (K)	0 (s)	0 ($1s$)
2 (L)	0 (s)	0 ($2s$)
	1 (p)	-1 ($2p_x$), 0 ($2p_z$), 1 ($2p_y$)
3 (M)	0 (s)	0 ($3s$)
	1 (p)	-1 ($3p_x$), 0 ($3p_z$), 1 ($3p_y$)
	2 (d)	-2, -1, 0, 1, 2; 5 orbitals

Fig. 9.2 Relations between n, l, m_l and shells, subshells, and orbitals.

(b) Orbitals with the same n but different l's form subshells of the given shell. Subshells are denoted as s, p, d, f, g, etc. pertaining to $l = 0, 1, 2, 3, 4$, etc. When $n = 1, l$ can have only one value, 0, and the subshell can have only one orbital $1s$. When $n = 2$, l can have the values 0 and 1, giving rise to the two subshells s and p. The number of orbitals of a subshell may be calculated from $2l + 1$. Thus, in the subshell $1s$ there is one orbital; in the subshell p there are 3 orbitals, etc.

Figure 9.2 shows relations between shells, subshells, and orbitals. In many atoms, except hydrogenic atoms, the increase in energy does not always correspond to an increase in n. For example, the $4s$ energy is lower than the $3d$ energy, etc.

9.6. Shapes of Orbitals

Shapes of orbitals play an important role in determining how atoms bind to form molecules.

Question: An s orbital has its maximum at the center, yet the probability of finding the electron there is zero. Why? Reason: The quantity $\psi^*\psi$ is maximum at the center, but it has to be multiplied by the volume element, which is zero at the center, $\delta V = 0$.

Radial Distribution Function
Most often, one is interested in the distance of an electron from the nucleus (regardless of angles) rather than in a given volume element. To obtain a working formula, note that the volume element is $\delta V = 4\pi r^2 dr$. The probability of finding the electron in that volume element is

$$\text{Prob.} = \psi^*\psi \, 4\pi r^2 dr \qquad (9.9)$$

The quantity $4\pi r^2\,\psi^*\psi$ is the *radial distribution function* [sometimes denoted as $g(r)$] and represents the probability of finding the electron between r and $r + dr$ in an atom.

Example 9.3.

(a) Calculate the radiation frequency, $\nu^*_{1\leftarrow 4}$, resulting from the transition of $n = 4$ to $n = 1$ in the Lyman Series of the spectrum of He^+.
(b) What is the probability of finding the He^+ electron within a volume element $\delta V = (1\,\mathrm{pm})^3$ at a distance $r = \frac{1}{2}a_o$?

Solution

(a) The atomic number of He is $Z = 2$, and so

$$\nu^*_{1\leftarrow 4} = Z^2 R_{\mathrm{H}}(1/1^2 - 1/4^2) = 4 \times 1.097 \times 10^5\,\mathrm{cm}^{-1}(1 - 1/16)$$
$$= 4.114 \times 10^5\,\mathrm{cm}^{-1}$$

(b)
$$\psi^2\delta V = (1/\pi)(Z/a_o)^3\exp(-2Zr/a_o)\delta V$$
$$= (1/\pi)(2/(5.292 \times 10^{-11}m))^3\exp\left[-2 \times 2 \times \frac{1}{2}(a_o/a_o)\right]$$
$$\times (10^{-12}m)^3$$
$$= 2.325 \times 10^{-6}$$

9.7. Electron Spin

This is the intrinsic angular momentum of the electron. It may be thought of as the electron spinning about its axis. Spin is a quantum mechanical concept and the above classical analogue should not be taken too seriously.

An electron spin is characterized by a spin quantum number, m_s, which can have the value 1/2 which means it spins in the clockwise direction, or $-1/2$, which means it spins in the anti-clockwise direction. If the spin quantum number is $m_s = 1/2$, the spin is often called α and denoted by an upward pointing arrow ↑. If $m_s = -1/2$ the electron spin is called β and denoted by an arrow pointing downward, ↓.

9.8. Structure, Transitions and Selection Rules

We have already mentioned that in a hydrogenic atom, the electron can have three quantum numbers n, l, and m_l. We must now also add m_s. Thus,

$$n = 1, 2, \ldots$$
$$l = 0, 1, \ldots, n - 1$$
$$m_l = 0, +1, -1, +2, -2, \ldots$$
$$m_s = +1/2, -1/2$$

Note that l cannot exceed $n - 1$, although it may be smaller.

Each energy level is n^2-fold degenerate, meaning that there are n^2 states which have the same energy. This rule is true only for hydrogenic atoms and does not apply to other atoms.

9.9. Many-Electron Atoms

The Schrödinger equation can only be solved exactly for hydrogenic atoms, yielding exact analytic expressions for the wave-functions. In all other cases, solutions are approximate.

As a first approximation, one can think of the wave-function of the atom as the product of the wave-functions of the individual electrons, i.e.

$$\Psi_{\text{atom}} = \psi(1)\psi(2)\psi(3)\ldots \tag{9.10}$$

where $\psi(1)$ is the orbital of electron 1, $\psi(2)$ the orbital of electron 2, etc. but with *nuclear charge* that is modified by the presence of all other electrons. This effective nuclear charge, Z_{eff}, is the charge of the nucleus shielded by the other electrons. Thus, the nuclear charge an electron "sees" is not the actual charge, Z_e (e being the absolute value of an electronic charge) but $Z_{e-\sigma} = Z_{\text{eff}}$, where σ is a shielding constant that can be approximated.

The Z_{eff} values are different for s, p, d, etc. orbitals. For example, the s electron has greater penetration to the nucleus than the p electron; the p electron has greater penetration than the d electron, etc. But there are exceptions; for example, $4s$ precedes $3d$.

9.10. Pauli Exclusion Principle

This Principle states that no more than two electrons can occupy the same orbital. Actually, the Pauli Exclusion Principle really states that in an atom no two electrons can have the same four quantum numbers. If two electrons are in the same orbital, they will have the same quantum numbers n, l, and m_l but must have different spin quantum numbers (i.e. opposite spin quantum numbers). Obviously, no other electron can be put in this orbital. If another electron is put in this orbital, two spin quantum numbers will have to be equal, and this violates Pauli's Principle.

Pauli's Principle forms the basis for the *Aufbau (Built-up) Principle*, and is essential for determining atomic structures by applying Hund's Rule. Hund's rule states that *"the most stable configuration favors unpaired electrons among degenerate orbitals"*.

Periodic Trends

Recall from freshman chemistry that the atomic radius *increases* down the group in the periodic table and *decreases* across the period. On the other hand, the ionization energy *decreases* down the group and *increases* across the period. This can be explained by observing that the orbital "radius" progresses as follows: $1s < 2s, 2p < 3s, 3p < 4s, 4p \ldots$

9.11. Selection Rules for Spectroscopic Transitions

These rules require that

$$\Delta l = 1 \text{ or } -1 \text{ (also that } \Delta m_l = 0, +1 \text{ or } -1) \qquad (9.11a)$$

and that

$$\Delta n = \text{unrestricted} \qquad (9.11b)$$

enabling one to determine which transitions are allowed and which are forbidden.

The reason behind these rules has to do with the *spin of a photon*, which is *one*. Thus, if an atomic electron jumps from a p orbital to an s orbital there is a loss of *one unit* of angular momentum. The emitted photon carries off this unit, and angular momentum is conserved.

Chapter 10

Quantum Theory. The Chemical Bond

Concepts developed in previous chapters, especially those involving orbitals, can be extended to describe electronic structures of molecules.

There are two quantum mechanical theories of electronic structure of molecules: Valence Bond Theory and Molecular Orbital Theory.

1. *Valence Bond Theory.* The starting point is the concept of shared electron pairs. The theory introduces the concepts of σ *and* π *bonds, promotion, hybridization.*
2. *Molecular Orbital Theory.* The idea is that there are orbitals (wavefunctions) that spread over all atoms in the molecule. As in atoms, the *Aufbau* Principle is used to determine the electron configuration of molecules.

Preceding the quantum mechanical description of molecular structures, it is well to mention treatments based on ideas introduced by G.N. Lewis in 1916. Among these was the *Octet Rule*, stating that an atom in a molecule likes to be surrounded by 8 electrons, normally denoted by dots. But there are exceptions to the octet rule — sometimes there are more than 8 electrons, sometimes less. Furthermore, the original theory could not account for the shape of molecules, but when coupled with VSEPR (Valence Shell Electron Pair Repulsion), the theory is remarkably successful in predicting shape.

Lewis already recognized before the advent of quantum theory that there are two major types of bonds: *covalent bonds* (in which electrons are shared) and *ionic bonds* (in which cohesion is due to electrostatic attraction between positive and negative ions).

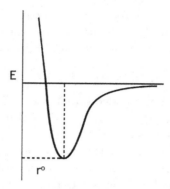

Fig. 10.1   Variation of potential energy of an electron with internuclear distance.

The only molecule that can be solved exactly is the *hydrogen molecule ion, H_2^+*. All other molecules must be solved approximately. An approximation that greatly simplifies quantum-mechanical treatments of molecules is the *Born–Oppenheimer Approximation*, which assumes that nuclei (being much heavier than electrons) move so slowly that they can be considered stationary. This approximation enables one to solve the Schrödinger equation and obtain the potential energy curves of an electron as a function of the internuclear distance (see Fig. 10.1).

10.1. Valence Bond Theory

The electrons that participate in bond formation are described by wave-functions that allow each electron to be on both atoms. For example, if Ψ_{1sA} and Ψ_{1sB} denote the orbitals on atom A and on atom B respectively, the wave-function of H_2 can be written

$$\psi_{1sA}(1)\psi_{1sB}(2) \tag{10.1a}$$

where (1) and (2) refer to electrons 1 and 2. An equally valid description is

$$\psi_{1sA}(2)\psi_{1sB}(1) \tag{10.1b}$$

A better description of the bond is a linear combination of both functions, yielding the two functions Ψ and Ψ^*

$$\Psi = \psi_{1sA}(1)\psi_{1sB}(2) + \psi_{1sA}(2)\psi_{1sB}(1) \tag{10.1c}$$

$$\Psi^* = \psi_{1sA}(1)\psi_{1sB}(2) - \psi_{1sA}(2)\psi_{1sB}(1) \tag{10.1d}$$

Neither of these functions represent correctly the H_2 molecule. What is missing are electron spin contributions. Denoting the spins as α (spin up) and β (spin down), we can construct four types of spin functions, namely

$$[\alpha_1\alpha_2], \quad [\beta_1\beta_2], \quad [\alpha_1\beta_2 + \beta_1\alpha_2], \quad [\alpha_1\beta_2 - \beta_1\alpha_2] \qquad (10.1e)$$

where the subscripts refer to electrons 1 and 2. Note that of the four spin functions [ ], the first three are symmetric with respect to exchange of the electrons (do not change sign); the fourth spin function is anti-symmetric.

There is an important rule that stipulates that for a wave-function to be proper, the function must change sign upon interchange of two electrons. Accordingly, the first three spin functions of Eq. (10.1a), being symmetric, can only combine with the orbital function Ψ^* given in (10.1a), which is anti-symmetric. The fourth spin function, which is anti-symmetric, can only combine with the symmetric function Ψ of (10.1c). Thus, there is one Ψ function and there are 3 Ψ^* functions. The state, characterized by Ψ, is called *singlet state* and the state characterized by Ψ^* is called *triplet state*.

Note: A linear combination (singlet or triplet) is the only formula which gives a correct description of combing atoms. It establishes that you cannot say whether electron 1 is on A and electron 2 on B or vice versa. In quantum mechanics, identical particles are indistinguishable, in contrast to classical mechanics where they are not. In general, a bond between A and B is

$$\Psi(A - B) = \psi_A(1)\psi_B(2) + \psi_A(2)\psi_B(1) \qquad (10.2)$$

Example 10.1.
(a) Describe the valence bond between two N atoms.

Solution
First note that the electron configuration of the N atom is $1s^2 2s^2 2p_x^1$ $2p_y^1 2p_z^1$. The $1s$ electrons are so deeply embedded that they are unlikely to take part in the bond formation and are ignored. The $2s$ electrons are also more deeply embedded than the p electrons and will also be ignored in bond formation. Only $2p$ electrons will be considered to participate in bond formation. It is common practice to take the Z-axis to be the internuclear axis. The $2pz$ orbitals of the two atoms point towards each other, while

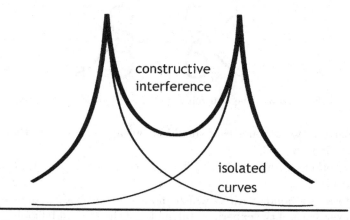

constructive
interference

isolated
curves

Fig. 10.2 Schematic representation of two overlapping atoms showing constructive interference.

the x and y orbital are perpendicular towards the bond axis. So, the wavefunction for the N_2 molecule is

$$\Psi(N - N) = \psi_{2p_zA}(1)\psi_{2p_zB}(2)$$
$$+ \psi_{2p_zA}(2)\psi_{2p_zB}(1) \qquad (10.3)$$

This form gives rise to constructive interference between the two components in the intermediate region (see Fig. 10.2).

(b) Describe the valence-bond ground state of Cl_2. The electron configuration of Cl is [Ne] $3s^2 3p_x^2 3p_y^2 3p_z^1$. The arguments are similar as for N_2 except that the bond orbitals are $3p_z$ orbitals.

$$\Psi(Cl - Cl) = \psi_{3p_zA}(1)\psi_{3p_zB}(2) + \psi_{3p_zA}(2)\psi_{3p_zB}(1) \qquad (10.4)$$

The bonds described so far are σ bonds. A σ bond has cylindrical symmetry along the internuclear axis. In the case of N_2, there are other orbitals of level 2, the p_x and p_y orbitals. These can overlap if the internuclear distance is short. They give rise to π orbitals.

10.2. Polyatomic Molecules

Consider the water molecule H_2O. The valence configuration of O is $2s^2 2p_x^2 2p_y^1 2p_z^1$. The two unpaired electrons can each pair with a H_{1s} orbital,

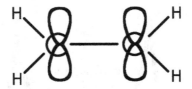

Fig. 10.3   Diagram representing sp^2 hybrids in $CH_2=CH_2$.

resulting in two σ bonds that should be 90° apart. However, the bond angle turns out to be 104°.

Another example is NH_3. The valence electron configuration of N, as noted before, has 3 unpaired electrons, namely $2p_x$, $2p_y$ and $2p_z$ and thus can form $3\,\sigma$ bonds with H which should be perpendicular to each other. However, each of the experimental angles of the pyramidal molecule is 107°.

Thus, the valence bond theory is not a good predictor of the bond angles. But where valence bond theory is particularly deficient is in predicting the shape of the molecule CH_4. The valence electron configuration of C is $2s^2 2p_y^1 2p_z^1$ and should have 2 bonds separated by 90°. But CH_4 has 4 bonds about 109° apart.

To overcome this deficiency, two additional concepts had to be introduced, the concept of *promotion* and the concept of *hybridization*. Specifically, in the case of CH_4, one of the $2s$ electrons is *promoted* to a $2p$ state, forming a $2p_x^1$ orbital, thus giving rise to the C configuration $2s^1 2p_x^1 2p_y^1 2p_z^1$. These orbitals form four *hybrid* orbitals, each consisting of the combination of $1s$ and $3p$'s. These orbitals are called sp^3 orbitals. Each of these hybrid orbitals consists of a small and a large lobe. The large lobes point towards the corners of a tetrahedron.

Other hybrid orbitals have been introduced to explain other molecules, for example ethylene, $CH_2=CH_2$ (Fig. 10.3). Ethylene makes use of sp^2 orbitals which are planar and 120° apart.

Sometimes d orbitals must be included to explain certain structures, for example, PCl_5. Even sp hybrids have been used to explain some structures, for example, $HC\equiv CH$.

10.3. Molecular Orbital Theory

It is assumed here that the electrons spread over the entire molecule, rather than belonging to a particular bond. The essential features of molecular orbital theory are illustrated by H_2^+ which has an exact solution.

To a good approximation, the wave-function of the hydrogen-molecule ion is

$$\Psi = \psi_{1sA} + \psi_{1sB} \tag{10.5}$$

where the two wave-functions on the right are respectively the orbitals centered on A and B and Ψ *is the molecular orbital.* This type of approximation is referred to as *linear combination of atomic orbitals* (*LCAO*) and the molecular orbital is referred to as *LCAO-MO*.

The molecular orbital in the above example is a σ-orbital because of its cylindrical symmetry. The probability density of the molecule can be written as

$$\Psi^2 = \psi_{1sA}^2 + \psi_{1sB}^2 + 2\psi_{1sA}\psi_{1sB} \tag{10.6}$$

Here,

- ψ_{1sA}^2 is the probability density of finding the electron on A.
- ψ_{1sB}^2 is the probability density of finding the electron on B.
- $2\psi_{1sA}\psi_{1sB}$ is the overlap density arising from the *constructive interference* of the atomic wave-functions. An electron can interact with both nuclei, binding them.

10.4. Bonding and Anti-bonding Orbitals

From the linear combination of atomic orbitals, one can construct two molecular orbitals, a bonding and an anti-bonding one.

$$\Psi = \psi_A + \psi_B \quad \text{bonding orbital} \tag{10.7a}$$

$$\Psi^* = \psi_A - \psi_B \quad \text{anti-bonding orbital} \tag{10.7b}$$

Bonding orbitals, if occupied, result in lowering the (electron) energy, as compared with the energy of the atom (causing an increase in cohesion). Anti-bonding orbitals have higher energy and cause a reduction in cohesion.

Examples of MO representation diagrams are shown in Fig. 10.4 and 10.5. Note that these diagrams show energy levels of both separated and combined atoms, depicting atomic and molecular orbitals. The principles on which these results are based are the same as for atomic orbitals, namely the *Aufbau Principle, Hund's Rule*, etc.

(Here, as customary, anti-bonding orbitals are denoted by *). Each MO can hold a maximum of two electrons having opposite spins. The π orbitals

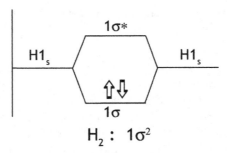

$$H_2 : 1\sigma^2$$

Fig. 10.4a   MO representation of H_2.

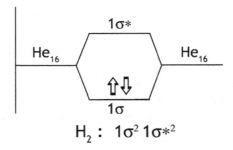

$$H_2 : 1\sigma^2\,1\sigma{*}^2$$

Fig. 10.4b   MO representation of He_2.

are degenerate, (depicted on the diagrams by double lines). Note that for molecules heavier than N_2 there is a reversal of the 1π and 2σ orbitals (Fig. 10.5).

Example 10.2. Write down the electronic configuration of the following molecules N_2, O_2, F_2, and Ne_2.

Solution
The electronic configurations are

$$N_2 : \ 1\sigma^2 \ 1\sigma^{*2}1\pi^4 2\sigma^2 \tag{10.8}$$

$$O_2 : \ 1\sigma^2 \ 1\sigma^{*2}2\sigma^2 1\pi^4 1\pi^{*2} \tag{10.9}$$

$$F_2 : \ 1\sigma^2 \ 1\sigma^{*2}2\sigma^2 1\pi^4 1\pi^{*4} \tag{10.10}$$

$$Ne_2 : \ 1\sigma^2 1\sigma^{*2}2\sigma^2 1\pi^4 1\pi^{*4} 2\sigma^{*2} \tag{10.11}$$

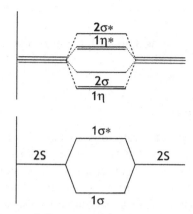

Fig. 10.5a MO diagram of diatomic molecules of atoms up to and including N.

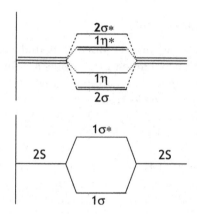

Fig. 10.5b Diagram of MO orbitals of diatomic atoms greater than N.

10.5. Bond Order

Often, the bonds are characterized by an *order*, which is the difference between the number of bonding and anti-bonding orbitals divided by 2. The greater the order is, the shorter the bond is. Thus, for N_2, O_2, F_2 and Ne_2, the bond orders are respectively 3, 2, 1, 0.

Example 10.3. Which molecule has the greater ionization energy:

(1) N_2 or N_2^+?   (2) F_2 or F_2^+?

Solution

(1) The bond order of N_2 is 3, as we have seen. The electronic configuration of N_2^+ is $1\sigma^2\ 1\sigma^{*2}\ 1\pi^4\ 2\sigma^1$ with bond order of 2.5. Accordingly, N_2 has larger ionization energy.

(2) The bond order of F_2 is 1 and of F_2^+ is 3/2. Thus, F_2^+ has greater ionization energy.

Homonuclear diatomic molecules are also characterized by *parity*. This is a concept that tells whether the wave-function changes sign or remains unchanged upon reflection through a center of symmetry. The σ orbitals have even parity (do not change sign) whereas the σ^* orbitals have odd parity (do change sign). The π orbitals have odd parity, whereas the π^* orbitals have even parity. It is common practice to designate the even and odd parities by the subscripts g and u. Thus we can write σ_g, σ_u^*, π_u, π_g^*.

10.6. Polar Covalent Molecules

Electron pairs between identical atoms are shared equally by the two atoms. But when the atoms are different, the atoms are pulled closer to one atom than the other. The ability of an atom to draw electrons close to it is called *electronegativity*. Here are electronegativity values of some light elements: H(2.1), Li(1.01), Be(1.5), B(2.0), C(2.5), N(3.0), O(3.5), F(4.0), Na(0.9).

Why do some atoms pull electrons closer to themselves than others? The rationale for this may be inferred by analyzing the HF molecule. The ionization energy of H is 13.6 eV; for F it is 18.8 eV. Thus, it would take about 5 eV more to pull off an electron from F than from H, that is the energy levels of F and H are not equal; the energy of F is $\sim$5 eV lower. The MO for a heterogeneous diatomic molecule must have different weighting factors for the atomic orbitals (Fig. 10.6). In the case of HF,

$$\Psi(\text{HF}) = c_{\text{H}}\psi(\text{H}) \pm c_{\text{F}}\psi(\text{F}) \tag{10.12}$$

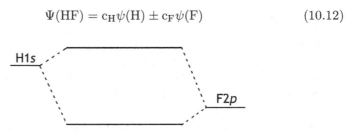

Fig. 10.6 Diagram of MO orbitals of diatomic molecules of unequal atoms.

where $c_H = 0.33$ and $c_F = 0.94$ in bonding and $c_H = 0.94$ and $c_F = 0.33$ in anti-bonding orbitals. Thus,

$$\Psi_{bnd} = 0.33\psi(H) + 0.94\psi(F) \tag{10.13a}$$

$$\Psi_{antb} = 0.94\psi(H) - 0.33\psi(F) \tag{10.13b}$$

10.7. Structure of Polyatomic Molecules

In the molecular orbital approach (LCAO-MO), each molecular orbital is a linear combination of all the atomic orbitals.

In benzene (C_6H_6), the six C atomic orbitals are sp^2 hybrids, forming σ bonds at 120° angles. The p orbitals form π bonds that spread over the entire molecule. The π bonds are delocalized and each pair helps to bind several atoms. Delocalization is an important feature of MO bonding.

10.8. Normalization. Normal Constants

The wave-functions of all systems should be normalized, meaning that the square wave-function should integrate to 1. For example,
   Particle-in-a-box:

$$\psi = N \sin n\pi x/L \tag{10.14}$$

Normalization, $N^2 \int_0^L dx \sin^2 n\pi/L = 1$, gives

$$N = \sqrt{2/L} \tag{10.15}$$

It is customary to include the normalization factor in atoms with the wave-function. Thus, the ground state wave-function for H, for example, is $\psi_{1s} = (1/\sqrt{\pi})(Z/a)^{\frac{3}{2}} e^{-(Z/a)r}$ which includes the normalization factor.

10.9. Normalization Molecules (MO)

Assuming that the atomic orbitals are already normalized, we can write

$$\Psi = \psi_A + \psi_B$$

$$N^2 \int \Psi^2 dv = N^2 \left[\int \psi_A^2 dv + \int \psi_B^2 dv + 2 \int \psi_A \psi_B dv \right]$$

$$= N^2[1 + 1 + 2S] = 1 \tag{10.16a}$$

where S is

$$S = \int \psi_A \psi_B dv \tag{10.16b}$$

called the *overlap integral*. Obviously, the molecular normalization coefficient is

$$N = 1/\sqrt{2(1+S)} \tag{10.16c}$$

A heteronuclear diatomic molecule has the wave-function

$$\Psi = \psi_{sA} + \lambda \Psi_{sB} \tag{10.17}$$

Normalization gives

$$N^2 \left[\int \psi_{sA}^2 dv + 2\lambda \int \psi_{sA} \psi_{sB} dv + \lambda^2 \int \psi_{sB}^2 \right] = 1$$

$$N^2[1 + 2\lambda S + \lambda^2] = 1$$

$$N = \{1/(1 + 2\lambda S + \lambda^2)\}^{1/2} \tag{10.18}$$

Chapter 11

Elements of Molecular Spectroscopy

11.1. Vibration–Rotation Spectra of Diatomic Molecules

Two points should be emphasized:

1. Actual energy levels are not directly measurable.
2. A spectrum shows only transitions of energy.

11.2. Rotational Selection Rules

Spectral absorption or emission corresponds to transitions between pairs of levels, as noted before. But not all possible levels may be combined. By theoretical studies of the wave-functions, especially their symmetry properties, one can devise selection rules. These rules tell whether certain quantum jumps are allowed or not allowed. Selection rules do not tell whether the intensities are weak or strong.

The rotational energy levels of a linear molecule are

$$E_J = J(J+1)(h/2\pi)^2/2I \quad J = 0, 1, 2, \ldots \tag{11.1}$$

where J is the rotational quantum number, and I is the moment of inertia, namely $I = \mu d^2$ (d being the internuclear distance and μ the reduced mass). The results are generally expressed in terms of the *rotational constant*, B, which has the dimensions of Hz (s^{-1}) and is defined as

$$B = (h/8\pi^2 I) \tag{11.2a}$$

or, in terms of B*, defined as

$$B^* = B/c \tag{11.2b}$$

which has the dimensions of cm^{-1} (c is the velocity of light). The energy of radiation of a transition from J' to J is $h\nu_{J \leftarrow J'} = E_{J'} - E_J$, yielding a transition frequency in s^{-1}

$$\nu_{J \leftarrow J'} = (E_{J'} - E_J)/h$$
$$= [J'(J' + 1) - J\{J + 1\}]h/(8\pi^2 I)$$
$$= [J'(J' + 1) - J\{J + 1\}]B \qquad (11.2c)$$

The transition frequency is often expressed in wave-numbers, $\nu^* = \nu/c$, yielding values in cm^{-1},

$$\nu^*_{J \leftarrow J'} = [J'(J' + 1) - J(J + 1)]B^* \qquad (11.3a)$$

where,

$$B^* = B/c = h/(8\pi^2 cI) \qquad (11.3b)$$

For pure rotation, the selection rule requires that ΔJ be either $+1$ or -1.

11.3. Vibrational Selection Rules

The vibrational energy of a diatomic molecule (simulated by a harmonic oscillator) is

$$E_v = (v + 1/2)h\nu_o \quad v = 1, 2, \ldots \qquad (11.4)$$

where v is the vibrational quantum number and ν_o is the vibration frequency, often expressed in terms of wave-numbers, $\nu^*_o = \nu_o/c$, which has the dimension of cm^{-1}. The energy can be expressed as

$$E_v = (v + 1/2)hc\nu^*_o \qquad (11.5)$$

and the transition frequency, in wave-numbers cm^{-1}, as

$$\nu_{v \leftarrow v'}(v) = (v' - v)\nu^*_0 \qquad (11.6)$$

The selection rule for harmonic oscillators requires that Δv be $+1$ or -1.

11.4. Further Requirements

The foregoing selection rules are necessary but not sufficient conditions:

(1) For rotational transitions with no change in vibrational quantum number, the molecule must have a *dipole-moment*.

(2) For a vibration transition, with or without accompanying rotational transitions, the molecule must have a *non-vanishing dipole moment derivative*. The dipole moment must not vanish upon vibration.

11.5. Pure Rotational Spectra

These spectra are observed in the far infrared and the microwave region. In particular, the transition frequencies for absorption and emission are respectively,

$$\nu^*_{J+1\leftarrow J} = (E_{J+1} - E_J)/hc \tag{11.7a}$$

$$= [(J+1)(J+2) - J(J+1)]B^* \tag{11.7b}$$

$$= 2(J+1)B^* \tag{11.7c}$$

$$\nu^*_{J-1\leftarrow J} = (E_{J-1} - E_J)/hc \tag{11.8a}$$

$$= [(J-1)(J) - J(J+1)]B^* \tag{11.8b}$$

$$= -2JB^* \tag{11.8c}$$

Note: Spacing between adjacent absorption peaks equals $2B^*$ (i.e. it is constant).

Exercise: The spacing between adjacent rotational lines in the far IR of HCl is $20.5\,\text{cm}^{-1}$. Calculate

(1) the moment of inertia,
(2) the equilibrium separation of the two nuclei (or bond-length).

11.6. Vibration–Rotation Spectra

Vibrational energy levels are generally widely separated, implying that at ordinary temperatures the molecules are in their ground state. It is then reasonable to assume that there is only one vibration transition, namely from $v = 0$ to $v = 1$.

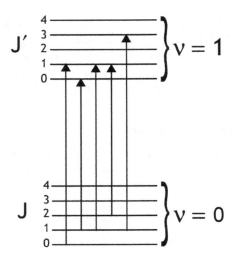

Fig. 11.1 Diagram representing transitions between $v = 0$ and $v = 1$ and $J \rightarrow J+1$ and $J \rightarrow J-1$.

Figure 11.1 depicts several rotational transitions between the vibrational level $v = 0$ and $v = 1$. Accordingly, there are two kinds (branches) of the spectrum:

$$v = 0 \rightarrow 1, \quad J \rightarrow J+1, \quad \text{R-branch} \tag{11.9}$$

$$v = 0 \rightarrow 1, \quad J \rightarrow J-1, \quad \text{P-branch} \tag{11.10}$$

The transition frequencies (in wave-numbers) are respectively,

$$\nu_R^* = \nu_o^* + 2B^* \left(J + 1 \right) \tag{11.11}$$

$$\nu_P^* = \nu_o^* - 2B^* J \tag{11.12}$$

Example 11.1. The vibration–rotation spectrum of ^{1}H ^{35}Cl has peaks at $\nu^*(\text{cm}^{-1})$: 2821.58, 2843.63, 2865.10, 2906.25, 2925.91, 2944.92 (see Fig. 11.2).

Note that

(1) The spacing between two adjacent levels in either the R- or P-branch is $\Delta\nu^* = 2B^*$.
(2) The lowest peak of the R-branch is obtained when $J = 0$ [Eq. (11.11)] and the highest peak of the P-branch is obtained when $J = 1$ [Eq. (11.12)]. The spacing between these two peaks is $4B^*$.

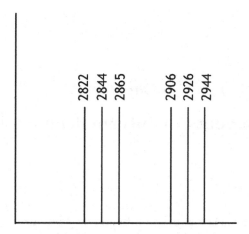

Fig. 11.2 Vibration–rotation spectrum of $^1\mathrm{H}^{35}\mathrm{Cl}$.

The spacing between all adjacent lines is the same except the spacing between the lines 2865.10 and 2906.25, which is twice as large. This gap separates the R- and P-branches. The above information suggests that

$$J = 2 \to 3 \quad \nu_R^* = \nu_o^* + 2B^* \times 3$$
$$\text{R} \quad J = 1 \to 2 \quad \nu_R^* = \nu_o^* + 2B^* \times 2 \tag{11.13a}$$
$$J = 0 \to 1 \quad \nu_R^* = \nu_o^* + 2B^* \times 1$$

$$J = 1 \to 0 \quad \nu_R^* = \nu_o^* - 2B^* \times 1$$
$$\text{P} \quad J = 2 \to 1 \quad \nu_R^* = \nu_o^* - 2B^* \times 2 \tag{11.13b}$$
$$J = 3 \to 2 \quad \nu_R^* = \nu_R^* - 2B^* \times 3$$

It is clear that by adding the lowest R peak and the highest P peak, and dividing the sum by 2 we can determine ν_0^*. Thus,

$$\nu_0^* = 1/2\,(2906.25 + 2865.10)\,\mathrm{cm}^{-1}$$
$$= 2995.6\,\mathrm{cm}^{-1} \tag{11.14}$$

This immediately gives $B^* = 10.34\,\mathrm{cm}^{-1}$, from which we can obtain the moment of inertia, I, and the internuclear distance, d.

Chapter 12

Elements of Intermolecular Forces

So far, the emphasis has been on bonds between atoms and between ions. The forces (interaction energies) can be dubbed *valence* forces. How do these simple molecules interact to form larger aggregates? This occurs via *intermolecular forces*. These forces account for the deviation from ideal behavior of gases, for formation of liquids and solids, for surface tension, viscosity and a host of other phenomena.

12.1. Types of Intermolecular Forces

12.1.1. Electrostatic Forces

12.1.1.1. Ion–Ion Forces

These take place between ions in solution. For example, NaCl dissolved in water contains Na^+ and Cl^- ions, which attract and repel each other. (The interaction energy varies as Q_1Q_2/R, where the Q's are the charges of the ions and R is the distance between them.)

12.1.1.2. Ion-Dipole Forces

These exist between an ion and the (partial) charge of a polar molecule (see Fig. 12.1). (The interaction energy varies as $Q\mu/R^2$, where Q is the charge and μ the dipole moment.)

12.1.1.3. Dipole–Dipole Forces

The molecules have dipole moments but are neutral. However, there is attraction between the positive end of one molecule and the negative end of another (Fig. 12.2). (The interaction energy varies as $\mu_1\mu_2/R^3$.)

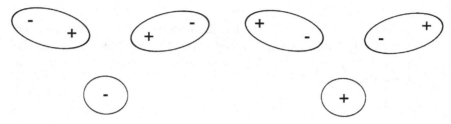

Fig. 12.1 Ion–dipole interaction.

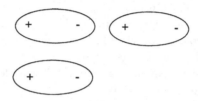

Fig. 12.2 Dipole–dipole interaction.

12.1.2. *van der Waals Forces*

12.1.2.1. *Orientation Forces*

These exist between rotating dipolar molecules that arise from the *correlative* rotational motion of the permanent dipole moments in neighboring molecules. The interaction energy has,

$$E_r = -\frac{2}{3}\mu_1^2\mu_2^2/(R^6kT) \tag{12.1}$$

The inverse R^6 accounts for the a/V^2 term in the van der Waals equation of state. Note that this interaction energy is temperature-dependent. The negative sign indicates that the interaction energy is attractive.

12.1.2.2. *Induction Forces*

These forces exist between a polar and another (polar or nonpolar) molecule. A polar molecule can induce a dipole moment in another molecule by distorting its charge distribution.

Comment: The ease with which a charge distribution can be distorted is called the *polarizability*, denoted as α. The induction energy has the form, $E_i = -\alpha_1\mu_2/R^{-6}$. The negative sign indicates that energy is attractive.

12.1.2.3. *London Dispersion Forces*

These forces exist between all kinds of molecular species (polar, nonpolar molecules, atoms, ions, etc.)

Even though nonpolar neutral molecules or atoms do not possess permanent dipole moments, they have instantaneous dipoles. The van der Waals or London dispersion forces arise from the correlative motion of these instantaneous dipole moments in neighboring molecules. For a diatomic molecule, the dispersion energy is

$$E_{dis} = -\frac{3}{2}\alpha_1\alpha_2 R^{-6} I_1 I_2/(I_1 + I_2) \qquad (12.2)$$

where I_1 and I_2 are the ionization energies of the two atoms. Note that the dispersion energy is temperature-independent.

Comments:

1) The van der Waals forces vary as R^{-6}, as noted before, resulting in V^{-2} dependent term of the equation of state.
2) The London forces are generally greater (stronger) than the orientation or induction forces, but weaker than the electrostatic forces.
3) Dispersion forces could be explained only after the advance of quantum mechanics. Classically, all motion ceases at $T = 0\,$K. In quantum mechanics there is motion, even at absolute zero.

12.2. Hydrogen Bonding

This exists between a H atom bonded covalently to either F, O or N and an unshared pair of electrons on a neighboring F, O or N. (Note: Hydrogen bondings are always the strongest of the intermolecular forces.)

Summary for neutral (non-ionic) molecules:

(1) London forces are always present, even in nonpolar molecules. If molecules are polar, orientation and induction forces are also preset in gases. The van der Waals forces are the weakest of the intermolecular forces.

(2) Dipole–dipole forces are present between polar molecules in liquids and in solids (where rotation is restricted or hindered). These forces are the next strongest of the intermolecular forces. Hydrogen bonding exists only between molecules containing F–H, O–H and N–H bonds. These are the strongest of the intermolecular forces.

Example 12.1.

1. What kind of intermolecular forces are present in the following substances?

   (a) CH_4,
   (b) chloroform ($CHCl_3$),
   (c) butanol ($CH_3CH_2CH_2CH_2OH$)

Solution

(a) CH_4 is nonpolar; hence, only London forces.

(b) $CHCl_3$ is unsymmetrical and has polar bonds. Therefore, in the liquid there are dipole–dipole and London forces.

(c) Butanol has H attached to O; it has hydrogen bonding, dipole–dipole and London forces.

12.3. Intermolecular Forces and Liquid Properties

Properties that are dependent on intermolecular forces are:

(a) *Vapor Pressure.* The ease or difficulty with which molecules leave the liquid depends on the strength of attraction to other molecules, thus depending on intermolecular forces. Weak intermolecular forces give high vapor pressure. Strong intermolecular forces give low vapor pressure.

(b) *Boiling Points.* This is the temperature at which the vapor pressure equals the atmospheric pressure. Thus, the higher the vapor pressure, the lower the boiling point is. In other words, the boiling point is highest for liquids with strong intermolecular forces (IF), and lowest for liquids with weak IF.

(c) *Surface Tension.* This is the energy needed to increase the surface area of liquids. To increase the surface, it is necessary to pull the molecules apart against the attractive forces. Thus, the surface tension increases with strength of IF.

(d) *Viscosity.* This is a measure of the resistance to flow. The stronger the inter molecular forces, the higher the viscosity.

12.4. Properties of Liquids

	M	BPt	V Pr (mmHg)	Surf Ts (Jm^{-2})	Visc $(kg\,m)$
H_2O	18	100	1.8×10^1	7.3×10^{-2}	1×10^{-3}
CO_2	44	−78.5	4.3×10^4	1.2×10^{-3}	7.1×10^{-5}
C_5H_{12}	72	36.2	4.4×10^2	1.6×10^{-2}	2.4×10^{-4}
$C_3H_8O_3$	92	290	1.6×10^{-4}	6.3×10^{-2}	1.5×10^0
$CHCl_3$	119	61.3	1.7×10^2	2.7×10^{-2}	5.8×10^{-4}
CCl_4	154	76.7	8.7×10^1	2.7×10^{-2}	9.7×10^{-4}
$CHBr_3$	253	149.5	3.0×10^0	4.2×10^{-2}	2.0×10^{-3}

Note: $C_3H_8O_3$ stands for glycerol. Its structure is

$$
\begin{array}{ccc}
H & H & H \\
| & | & | \\
H-C-C-C-H \\
| & | & | \\
O & O & O \\
| & | & | \\
H & H & H
\end{array}
$$

Notes:

1) Except for H_2O and glycerol, the interactions are due to the London forces, and in a few cases to dipole–dipole forces. The liquids are listed in order of increasing molecular weight.

2) London forces vary as α^2, which increase with molecular weight. Thus, with exception of water and glycerol, there should be a decrease in vapor pressure going down the list, and there is. Except

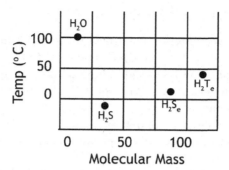

Fig. 12.3 Variation of the boiling point of hydrogen-bonded molecules with molecular weight.

for H_2O and $C_3H_8O_3$, the London force is dominant and increases going down the list. Therefore the boiling point should increase, and it does.

3) Surface tension should increase with increasing weight, and it does.
4) Viscosity should increase with increasing molecular weight and it does. (It should be mentioned that there are other factors involved, such as entanglements, but these are not discussed here.)

Figure 12.3 depicts the boiling point variation of several hydrogen-containing compounds with molecular weight. Note that the boiling point increases with increasing weight (larger polarizabilities, stronger London forces) except for water. In water, there are in addition to the London forces dipole–dipole forces and hydrogen bonding. Water has by far the strongest intermolecular forces and should have a much higher boiling point. The diagram clearly shows this.

12.5. Classification of Solids by Types of Intermolecular Forces

All solids consist of structural units (atoms, molecules, ions) which are strongly attracted to one another to give a rigid substance. In some cases, the forces binding the structural units are *chemical* bonds (they can be metallic, ionic, or covalent bonds). These bonds hold atoms together. In other cases, the structural units are molecules held together by intermolecular forces.

There are four different kinds of solids:

1. *Molecular Solid.* This is a solid held together by intermolecular forces. Examples are solid Ne, solid H_2O, and solid CO_2

2. *Metallic Solid.* This is a solid composed of a positive charged core of atoms, held together by a "sea" of delocalized electrons. Examples are iron, copper, and silver.

3. *Ionic Solid.* This is a solid that consists of cations and anions held together by the attraction of opposite charges (ionic bonds). Examples are CsCl and NaCl.

4. *Covalent Solid.* This is a solid held together by a large network or chain of covalent bonds. Examples are diamond and graphite sheets,

Appendix A

Homework Problem Sets

Problem Set I. Chapter 1

1. Calculate the pressure of 0.300 g of Ar occupying a volume of 400 dm^3 at 450 K.

2. A gas in a constant volume container has a pressure of 150 kPa at 25°C. Calculate the pressure at 250°C.

3. A gas in a balloon has a pressure of 100 kPa, a temperature of 25°C and a volume of 10 cm^3. What will the volume of the balloon be if it rises to an altitude where the atmospheric pressure is 100 Pa and the temperature is −50°C?

4. The balloon of Problem 3 is lowered at sea to a depth of 30 m. The density of sea water is 1.025 g/cm^3 and assumed to be constant. The temperature is also assumed to be constant. The gravitational constant is $g = 9.81\,\mathrm{ms}^{-2}$. The atmospheric pressure is 100 kPa. What is the volume of the air in the balloon at a depth of 30 m?

5. A gas mixture is composed of 300 mg of CH$_4$ and 200 mg of N$_2$. The partial pressure of CH$_4$ is 120 kPa at 300 K. Calculate the partial pressure of N$_2$, the total pressure of the mixture and the volume of the mixture.

6. Calculate the molar mass of 40 mg of a gas, occupying a volume of 250 cm^3 at a pressure of 175 atm and a temperature of 300 K.

7. Calculate the root-mean-square speed, c_{rms} , of an Ar atom at a pressure of 1 bar and a temperature of 298 K.

8. The cross-section of Ar is $\sigma = 0.36\,\mathrm{nm}^2$. Calculate the root-mean-square speed, c_{rms}, the diameter, d, the mean free path, λ, and collision frequency, z. The pressure of the gas is 1 bar (10^5 Pa) and the temperature is 25°C.

9. Develop the van der Waals equation of state in virial form and obtain explicit expressions for the virial coefficients B and C in terms of the van der Waals parameters a and b. [Note: The expansion of $1/(1-x) \approx 1 + x + x^2 + \cdots$]

Problem Set II. Chapter 2

1. How much work is required to raise an object weighing $25\,\mathrm{kg}$ to a height of $5\,\mathrm{m}$ on the surface of the earth? (The gravitational constant is $9.81\,\mathrm{ms^{-2}}$.)

2. 5 moles of a gas is expanded isothermally and reversibly at $298.5\,\mathrm{K}$ from $100\,\mathrm{cm^3}$ to $300\,\mathrm{cm^3}$. Calculate the work.

3. The heat of fusion of Na is $27.4\,\mathrm{cal/g}$ at $370.95\,\mathrm{K}$. How many kilojoules of heat are required to melt $250\,\mathrm{kg}$ of Na?

4. A room of volume $110\,\mathrm{m^3}$ contains a gas (air), whose molar heat capacity at constant pressure is $C_{p,m} = 21.0\,\mathrm{JK^{-1}}$. If the room is heated from $300\,\mathrm{K}$ to $310\,\mathrm{K}$ (at constant volume), calculate the heat required to accomplish this. The molar volume of the gas at $T = 300\,\mathrm{K}$ is $25\,\mathrm{L\,mol^{-1}}$.

5. One mol of O_2 is heated at constant pressure from $250\,\mathrm{K}$ to $275\,\mathrm{K}$. The molar heat capacity at constant pressure is $C_{p,m} = 29.4\,\mathrm{JK^{1}\,mol^{-1}}$. Calculate $C_{V,m}, q_P, q_V, \Delta H, \Delta U$.

6. The standard enthalpy of formation of $CH_3C_6H_5$ (methyl benzene) is $+50.0\,\mathrm{kJ\,mol^{-1}}$. What is the standard enthalpy of combustion? (The standard enthalpies of formation of $CO_2(g)$ and of $H_2O(l)$ are respectively $-393.51\,\mathrm{kJ\,mol^{-1}}$ and $-285.83\,\mathrm{kJ\,mol^{-1}}$.)

7. The standard enthalpy of combustion of $C_8H_{18}(l)$ (octane) is $-546.1\,\mathrm{kJ\,mol^{-1}}$. What is the standard enthalpy of formation?

8. The standard enthalpy of formation of $AgCl(s)$ is $-127.1\,\mathrm{kJ\,mol^{-1}}$. The standard enthalpy of formation of $Ag^+(aq)$ is $105.6\,\mathrm{kJ\,mol^{-1}}$ and of $Cl^-(aq)$ it is $-167.2\,\mathrm{kJ\,mol^{-1}}$. Calculate the standard enthalpy of solution of AgCl in H_2O.

9. The enthalpy of combustion of diamond is $-395.4\,\mathrm{kJ\,mol^{-1}}$ and of graphite is $-393.5\,\mathrm{kJ\,mol^{-1}}$. Calculate the enthalpy of conversion of graphite into diamond.

10. $1.6\,\mathrm{g}$ of glucose ($M = 180.2\,\mathrm{g\,mol^{-1}}$.) burns in air (combines with oxygen) at constant pressure and temperature. The standard enthalpy of combustion of sucrose is $-2808\,\mathrm{kJ\,mol^{-1}}$. How much heat is released?

11. The standard enthalpy of combustion of 1-butene (C_4H_8, g) is $-2710\,kJ\,mol^{-1}$ and the standard enthalpy of vaporization of liquid 1-butene is $18\,kJ\,mol^{-1}$. Calculate the standard enthalpy and the standard energy of combustion of liquid 1-butene.

Problem Set III. Chapters 3, 4, 5

1. $250\,kg$ of Al is cooled from $298.15\,K$ to $200\,K$ at constant pressure. The molar heat capacity of Al is $24.4\,JK^{-1}mol^{-1}$. Calculate the heat emitted and the change in entropy.

2. $20\,L$ of CO_2 at $275\,K$ and 1 atmosphere pressure is compressed to a volume V_f. The change in entropy is $-12.0\,JK^{-1}$. Calculate V_f.

3. The enthalpy of vaporization of a substance is $30\,kJ\,mol^{-1}$ at the normal boiling point $T_b = 350\,K$. What is the entropy of vaporization of the substance, and what is the entropy change of the surroundings?

4. Calculate the standard reaction enthalpies for

   (a) $Hg(l) + Cl_2(g) \rightarrow HgCl_2(s)$
   (b) $Zn(s) + Cu^{2+}(aq) \rightarrow Zn^{2+} + Cu(s)$

5. The enthalpy of combustion of $C_6H_5OH(s)$ at $298\,K$ is $-3055\,kJ\,mol^{-1}$. The standard entropy of this compound at $298\,K$ is $146\,JK^{-1}\,mol^{-1}$. Calculate the

   (a) enthalpy of formation of $C_6H_5OH(s)$
   (b) entropy of formation of $C_6H_5OH(s)$
   (c) Gibbs free energy of formation of $C_6H_5OH(s)$
   (d) Gibbs free energy of the combustion reaction
   (e) the non-PV work, w_{other}

6. Assuming that at $2000\,K$ and $100\,kbar$ the transition reaction

$$C\ (\text{graphite}) \rightleftarrows C\ (\text{diamond})$$

   is reversible (in equilibrium), calculate the ΔS for this reaction. The standard enthalpy change at this temperature and pressure is $\Delta H^o_{reac} = 1.91\,kJ\,mol^{-1}$.

7. The standard Gibbs free energy of transformation of trans-2-pentane to cis-2-pentane at $127°C$ is $\Delta G^{\underline{o}} = 3.7\,kJ\,mol^{-1}$. Calculate the equilibrium constant for the isomerization of cis-2-pentane to trans-2-pentane.

8. The standard Gibbs free energy for a given reaction is $\Delta G^{\underline{o}} = 40\,\text{kJ}\,\text{mol}^{-1}$ at 1500 K. The enthalpy change for the reaction is $\Delta H = 250\,\text{kJ}\,\text{mol}^{-1}$ at 1500 K and approximately constant.

   (a) Calculate the equilibrium constant at $T = 1500\,\text{K}$.
   (b) Calculate the temperature at which the equilibrium constant is $K = 1$.

9. Which of the following reactions has a $K > 1$?

   (a) $NH_4Cl \rightarrow HCl(g) + NH_3(g)$
   (b) $Fe(s) + 2H_2S(g) \rightarrow FeS_2(s) + 2H_2(g)$
   (c) $H_2SO_4(l) + 2H_2(g) \rightarrow 2H_2O + H_2S(g)$

10. Which of the reactions in Problem 9 will be enhanced (yield more products) by increasing the temperature?

Problem Set IV. Chapter 6

1. A liquid in equilibrium with its vapor consists of substances A, B and C. If the volume of the vapor phase is 100 m^3 at 298.15 K and the partial pressures of the three substances are respectively, $P_a(A) = 2.33\,\text{kPa}$, $P_a(B) = 1.43\,\text{kPa}$ and $P_a(C) = 0.26\,\text{kPa}$, calculate

   (a) the mole fraction of each substance in the vapor phase,
   (b) the number of moles of each substance in the vapor phase.

2. The partial molar volumes of benzene $[C_6H_6(l)]$ and cyclohexane $[(C_6H_{12}(l)]$ are respectively 75 and 80 cm^3 mol^{-1}. The mole fraction of benzene is 0.6.

   (a) What is the mole fraction of cyclohexane?
   (b) What is the weight per mole of the combined substances?
   (c) What is the volume of a solution whose total mass is 500 g?

3. The mole fraction, x_B, and partial pressure, P_B, of three solutions of a substance is

x_B	0.010	0.024	0.038
P_B/kPa	64.0	153.8	243.6

   Calculate Henry's Law constant, K_H.

4. (a) The partial vapor pressure of oxygen $[O_2(g)]$ in air at sea level is 21.0 kPa at 298.15 K. Henry's Law constant at that temperature

is $K_H = 74.68\,\mathrm{kPa\,m^3\,mol^{-1}}$. What is the solubility (molarity) of oxygen in water at that temperature?

(b) If the minimum concentration of $O_2(g)$ in water necessary to sustain life is $4\,\mathrm{mg/L}$, what is the minimum partial pressure of $O_2(g)$?

5. The vapor pressure of a pure solvent at $80.0°C$ is 400 Torr. When 0.15 g of a solute is dissolved in 6.5 g of this solvent, the pressure is reduced by 15 Torr. The molar mass of the solvent is $M_{\text{solvent}} = 80.0\,\mathrm{g\,mol^{-1}}$. What is the molar mass, M_{solute} of the compound?

6. The addition of 30.0 g of a substance to 7.0 g of CCl_4 lowered the freezing point of the solvent by 7.5 K. The freezing point lowering constant is $K_f = 40\,\mathrm{K/mol}$ per 1 kg of solvent. What is the molar mass of the compound?

7. The osmotic pressure of an aqueous solution of a solid in $H_2O(l)$ is 125 kPa at 298.15 K. The mass number of the solid is $M = 180.16\,\mathrm{g\,mol^{-1}}$. The freezing point constant for water is $K_f = 1.86\,\mathrm{K\,kg_{solvent}/mol_{solute}}$. The normal freezing point of $H_2O(l)$ is $T_f^* = 273.15\,\mathrm{K}$. Calculate the freezing point of the solution.

8. Calculate the number of moles (the solubility) of CO_2 in water at 298.15 K when the partial pressure of CO_2 is

(a) $2.0\,\mathrm{kPa}$,
(b) $200\,\mathrm{kPa}$.

Henry's Law constant of CO_2 in water at 298.15 K is $K_H = 2.937\,\mathrm{kPa\,m^3\,mol^{-1}}$ and the atmospheric pressure is 1 atm or 101 kPa at 298.15 K.

9. Calculate the boiling point of $250\,\mathrm{cm^3}$ of water containing 7.5 g of sucrose. The normal boiling point of water is 373.15 K and the boiling point elevation constant is $K_b = 0.51\,\mathrm{K\,kg_{solv}\,mol_{solute}^{-1}}$.

10. (a) Solute B and solvent A form an ideal dilute solution. The partial pressure of the solute is 30 Torr when the mole fraction of the solute is $x_A = 0.20$. Find Henry's Law constant K_B.

(b) If the partial pressure of the solvent at that composition is 250 Torr, what is the pressure of the pure solvent?

11. An aqueous solution containing a substance whose concentration is $7.50\,\mathrm{g/L}$ has an osmotic pressure of 1150 Torr at 77 K.

(a) What is the molecular weight of the solute?

(b) What is the osmotic pressure of a solution at $27°C$ that contains 1.00 g of this solute in 500 mL of solution?

12. $NH_4Cl(s)$ dissociates in accordance with the reaction

$$NH_4Cl(s) \rightleftarrows NH_3(g) + HCl(g)$$

The total pressure at $700\,K$ is $6.00 \times 10^3\,Pa$ and at $730\,K$ it is $11.00 \times 10^3\,Pa$. The standard state pressure is $P^\circ = 101.2 \times 10^5\,Pa$.

(a) Calculate the equilibrium constant K at $T = 700\,K$ and the equilibrium constant K' at $T' = 730\,K$.

(b) Calculate the standard enthalpy change, $\Delta H^{\underline{o}}$, for this reaction. [Assume that within this temperature range $\Delta H^{\underline{o}}$ is constant]

13. Calculate the standard potentials for the cell reactions

(a) $Fe(s) + Pb^{2+}(aq) \rightarrow Fe^{2+}(aq) + Pb(s)$
(b) $H_2(g) + I_2(g) \rightarrow 2HI(aq)$

14. Calculate the equilibrium constants at $298\,K$ for the reactions

(a) $Sn(s) + Sn^{4+}(aq) \rightarrow 2Sn^{2+}(aq)$
(b) $Fe(s) + Hg^{2+}(aq) \rightarrow Hg(l) + Fe^{2+}(aq)$
(c) $Cu^{2+}(aq) + Cu(s) \rightarrow 2Cu^{+}(aq)$

Problem Set V. Chapter 7

1. The rate of formation of C in the reaction $3A + B \rightarrow 2C + D$ is $2\,mol\,L^{-1}\,s^{-1}$. What is the rate of formation of D? What is the rate of disappearance of A?

2. The reaction of Problem 1 was found to obey the reaction law $r = k[A][B][D]$. What are the units of k?

3. The rate of reaction $2N_2O_5(g) \rightarrow 4NO_2(g) + O_2$ is first order in N_2O_5, i.e. $r = k_A[N_2O_5]$. The rate constant $k_A = 3.46 \times 10^{-5}\,s^{-1}$ at $298\,K$.

(a) Calculate $t_{1/2}$ for this reaction

(b) If the initial pressure of N_2O_5 is $75.0\,kPa$, what is the total pressure of the gases?

  1. after 10 seconds?
  2. after 1 hour?

4. The oxidation of ethanol is first order in C_2H_5OH. If the concentration of ethanol decreased from $440\,mmol\,L^{-1}$ to $112\,mmol\,L^{-1}$ in $1.22 \times 10^4\,s$, calculate the rate constant of the reaction.

5. The concentration of a reactant in a second-order reaction dropped from $440\,\mathrm{mmol\,L^{-1}}$ to $112\,\mathrm{mmol\,L^{-1}}$ in $2.44 \times 10^4\,\mathrm{s}$. Calculate the rate constant of this reaction.

6. The decomposition of ammonia is a zero-order reaction in NH_3. The partial pressure of NH_3 drops from 20 to $10\,\mathrm{kPa}$ in $750\,\mathrm{s}$.

   (a) What is the rate constant for the reaction?
   (b) How long will it take for all ammonia to disappear?

7. The half-life of a substance is $250\,\mathrm{s}$. The reaction is first order. How long will it take for the substance to reduce to 10% of its initial value?

8. ^{14}C decays with a half-life of 5750 years. The reaction is of first order. If the percentage of ^{14}C in a skeleton contains 90% of ^{14}C that is present in living matter, how old is the skeleton?

9. The reaction $2A \rightarrow P$ is of second order, with $r = k[A]^2$. If $k_A = 0.75\,\mathrm{L\,mol^{-1}\,s^{-1}}$, how long would it take for the concentration of A to change from $0.40\,\mathrm{mol\,L^{-1}}$ to $0.02\,\mathrm{mol\,L^{-1}}$?

10. The concentration of B in the reaction $A \rightarrow 2B$ was measured at $20\,\mathrm{min}$ intervals and gave the following results:

$[B]/\mathrm{mol\,L^{-1}}$	0	0.178	0.306	0.400	0.460	0.624
t/min	0	20	40	60	80	∞

   What is the order of the reaction and what is the value of the rate constant, k?

11. For a given reaction, the rate constant is
    $k' = 2.0 \times 10^{-4}\,\mathrm{L\,mol^{-1}\,s^{-1}}$ at $T' = 298\,\mathrm{K}$, and
    $k = 4.5 \times 10^{-3}\,\mathrm{L\,mol^{-1}\,s^{-1}}$ at $T = 345\,\mathrm{K}$.
    Calculate the Arrhenius parameter A.

12. The activation energy of a given decomposition reaction is $E_a = 125\,\mathrm{kJ\,mol^{-1}}$. Estimate the temperature at which the reaction rate would be 15% greater than the rate at $300\,\mathrm{K}$.

13. Food decomposes (rots) 50 times faster at $30°C$ than at $0°C$. Estimate the activation energy of the decomposition.

14. The activation energy of the first-order decomposition of N_2O into N_2 and O is $E_a = 250\,\mathrm{kJ/mol}$. If the half-life of the reactant is $t_{1/2} = 7.5 \times 10^6\,\mathrm{s}$ at $673\,\mathrm{K}$, what will it be at $773\,\mathrm{K}$?

15. The mechanism for the reaction (catalyzed by Br^-)

$$2H_2O_2(aq) \rightarrow 2H_2O(l) + O_2(g)$$

obeys the following rules

$$H_2O_2(aq) + Br^-(aq) \rightarrow H_2O(l) + BrO^- \qquad \text{(slow)}$$
$$BrO^-(aq) + H_2O_2(aq) \rightarrow H_2O(l) + O_2(g) + Br^-(aq) \qquad \text{(fast)}$$

Determine the reaction rate, r, and the order of the reaction with respect to H_2O_2 and with respect to Br^-.

16. A reaction is found to obey the following mechanism.

$$A_2 \rightleftarrows 2A \qquad \text{(rapid equil.)}$$
$$A + B \rightarrow P \qquad \text{(slow)}$$

What is the rate law for the formation of P?

17. The mechanism for a chain reaction involving free radicals obeys the following rules:

   i. $[MN] \xrightarrow{k_a} [M\bullet] + [N\bullet]$

   ii. $[M\bullet] \xrightarrow{k_b} [O\bullet] + [Q]$

   iii. $[MN] + [O\bullet] \xrightarrow{k_c} [M\bullet] + [R]$

   iv. $[M\bullet] + [O\bullet] \xrightarrow{k_d} [P]$

The k_a, k_b, k_c, k_d are the reaction constants of the chain reactions. What is the reaction order with respect to [MN]?

Problem Set VI. Chapters 8, 9

1. If the wavelength of an electron is 500×10^{-12} m, what is its velocity? [The electron mass is $m_e = 9.1094 \times 10^{-12}$ kg; Planck's Constant is $h = 6.626 \times 10^{-34}$ J s; $1 J = 1 \text{ kg m}^2 \text{ s}^{-2}$.]

2. A photon knocks an electron out of an atom. If the ionization potential of the atom is 2.5×10^{-18} J and the speed of the electron is 10^6 ms^{-1}, what is the frequency of the incident photon?

3. If the uncertainty in position of a particle moving with a momentum of 5.0×10^{-20} kg m s^{-1} is $\Delta x = 9.0 \times 10^{-10}$ m, what fraction is the uncertainty in speed, Δp, of the actual momentum of the particle?

4. A hydrogen atom in a one-dimensional box of length $L = 10^{-9}$ m jumps from level 3 to level 2. How much energy is released?

5. An electron in a one-dimensional box of length $L = 5.0 \times 10^{-9}$ makes a transition from level 5 to level 3.

   (a) What is the energy release for this process?
   (b) What is the wavelength of the emitted radiation?

6. What is the energy of radiation corresponding to the following wavelength: (a) 650 nm (red), (b) 450 nm (violet), (c) 150 pm (X-rays)?

7. The work function of a metal is 2.0 eV. If the metal is subjected to radiation of wavelength (a) 250 nm, (b) 700 nm, will an electron be emitted, and if so, what will be the kinetic energy of the emitted electron?

8. Calculate the de Broglie wavelength of

   (a) a particle of mass of 1 g traveling with a speed of 2.0 ms^{-1},
   (b) a He atom traveling with a speed of 10^3 ms^{-1}.

9. What is the frequency of the transition of line $n = 6$

   (a) in the Balmer series?
   (b) in the Paschen series?

10. The probability of finding the electron in the H-atom in a small volume element at a radius r is 40% of its maximum value. Calculate the radius r.

11. Which of the following electronic transitions in an atom are allowed? which are forbidden? Why?

   (a) 3s → 2s, (b) 2p → 1s, (c) 4d → 3p, (d) 3d → 2s, (e) 5p → 2s.

12. How many electrons can occupy a subshell whose value is

   (a) l = 0, (b) l = 2, (c) l = 4. Explain.

13. The wavelength $\lambda = 6.577 \times 10^{-5}$ cm is observed in the spectrum of H. Show that this is a transition line in the Balmer Series. What is the wavelength of the next higher line?

Problem Set VII. Chapters 10, 11, 12

1. Give the ground state electron configuration of

   (a) H_2, (b) Li_2, (c) C_2, (d) N_2, (e) O_2, (f) F_2.

2. Which of the molecules in Problems 1 would you expect to be stabilized by

   (a) the addition of an electron?
   (b) the removal of an electron? Explain.

3. Arrange in accordance with increasing or decreasing bond length:

$$\text{(a) } F_2^-, F_2, F_2^+; \text{ (b) } N_2^-, N_2, N_2^+.$$

Give reason(s) for your choice.

4. In the FEMO (free electron molecular orbital) theory of conjugation, free electrons are treated as particles in a box. In butadiene there are 4 free electrons occupying the 2 lowest energy levels of the box. Taking the size of the box to be 560×10^{-12} m, calculate the excitation energy from the highest occupied level to the lowest unoccupied level.

5. The vibration–rotation spectrum of $^1H^{35}Cl$ has peaks at 2821.58, 2843.63, 2865.10, 2906.25, 2925.91, 2944.92 cm^{-1}. Calculate

   (a) the rotation constant B^*,
   (b) the vibration frequency v_o^*.

6. If the ^1H in Problem 5 were replaced by ^2H (deuterium), which molecule, ^1HCl or ^2HCl, will have the greater

   (a) vibration frequency,
   (b) moment of inertia,
   (c) rotational constant?

   Explain.

7. The spacing between successive rotational lines in the far infrared of $^1H^{35}Cl$ is 20.5 cm^{-1}. Calculate

   (a) the moment of inertia,
   (b) the bond length.

8. What kind of intermolecular forces are present in the

   (a) C_2H_6, (b) $CHCl_3$, (c) CH_3COH?

9. The boiling points of several hydrogen-containing compounds increase as follows:

$$H_2S < H_2Se < H_2Te < H_2O$$

Note that the heavier the molecule is, the higher the boiling temperature, except for H_2O, which is the lightest of these substances. Explain.

Appendix B

Thermodynamic Data

Table B.1. Standard enthalpies and free energies of formation in $kJ\,mol^{-1}$ and molar entropies in $J\,mol^{-1}$, all at $T = 298.15\,K$.

	$\Delta H_f^{\ominus}$ (kJ)	$\Delta G_f^{\ominus}$ (kJ)	$S_m^{\ominus}$ (J)
Ag^+(aq.)	105.56	−77.11	73.9
Al(s)	0	0	28.32
Al^{3+}(aq.)	−531.36	−485.34	321.75
Ar(g)	0	0	154.72
AgCl	−127.02	−109.70	96.11
AgBr(s)	−100.37	−96.90	107.11
Br_2(l)	0	0	152.21
C(graphite)	0	0	5.69
C(diamond)	1.88	2.89	1.59
CH_4(g)	−74.74	−50.79	186.18
C_2H_6(g)	−84.68	−32.88	229.49
CH_3COOH(l)	−487.02	−219.17	159.02
CH_3CHO(g)	−166.19	−128.88	250.33
C_2H_5OH(l)	−277.65	−187.32	160.67
C_3H_8(g)	−103.85	−23.49	209.91
C_6H_6(l)	49.64	167.36	202.92
$C_{10}H_8$(s)	78.53		
C_6H_5OH(s)	−165.0	−50.0	146.0
$C_{12}H_{22}O_{11}$(s)	−248.1	−156.4	200.4
CH_3OH(l)	−238.66	−166.81	126.72
$CHCl_3$(l)	−134.47	−73.72	201.67
CCl_4(l)	−139.49	−68.74	214.43
Cl_2(g)	0	0	223.09
Cl^-(aq.)	−167.16	−131.32	56.5
CO_2(g)	−393.51	−394.36	213.74
Cu(s)	0	0	33.15
Cu^{2+}(aq.)	64.77	65.49	−99.8
Fe(s)			27.28
Fe_2O_3(s)	−822.16	−740.99	84.96

(*Continued*)

Table B.1.   (*Continued*)

	$\Delta H_f^{\underline{O}}(kJ)$	$\Delta G_f^{\underline{O}}(kJ)$	$S_m^{\underline{O}}(J)$
$FeS_2(s)$	−177.91	−166.69	53.14
$FeSO_4(s)$	−928.43	−820.90	100.58
$H_2(g)$	0	0	130.58
$H_2O(l)$	−285.86	−237.19	69.96
$H_2O_2(l)$	−187.78	−120.41	109.62
$H_2S(g)$	−10.63	−33.56	206.79
$H_2SO_4(l)$	−813.99	−690.11	156.90
$HNO_3(l)$	−174.05	−80.79	155.60
$HCl(g)$	−92.29	−95.27	186.69
$H_3PO_4(l)$	−1279.05	−1119.22	110.49
$Hg(l)$	0	0	76.02
$HgCl_2(s)$	−224.3	−178.8	146.0
$I_2(s)$	0	0	116.15
$LiF(s)$	−612.12	−584.09	35.98
$K_2O(s)$	−361.49	−322.17	98.52
$Na(s)$	0	0	51.21
$NH_3(g)$	−46.19	−16.61	192.50
$NH_4Cl(s)$	−315.38	−208.89	94.56
$N_2O(g)$	82.05	104.20	219.85
$O_2(g)$	0	0	205.48
$Na_2O(s)$	−415.89	−376.66	72.89
$P(white)$	0	0	41.09
$P(red)$	−17.57	−12.13	22.81
$SO_2(g)$	−296.96	−300.07	248.52
$Zn(s)$	0	0	41.63
$Zn^{2+*}(aq.)$	−153.89	−147.08	−112.1
$ZnS(s)$	−205.98	−201.29	111.29

Appendix C

Standard Reduction Potentials

Table C.1. The standard (reduction) half-reaction potentials at $25°C$ for a number of ions.

	E^*/V
$Cd^{2+} + 2e^- \rightarrow Cd$	-0.40
$Cu^+ + e^- \rightarrow Cu$	$+0.52$
$Cu^{2+} + 2e^- \rightarrow Cu$	$+0.34$
$Fe^{2+} + 2e^- \rightarrow$	-0.44
$2H^+ + 2e^- \rightarrow H_2$	0
$2H_2O + 2e^- \rightarrow H_2 + 2OH^-$	-0.83
$Hg_2Cl_2 + 2e^- \rightarrow 2Hg + 2Cl^-$	$+0.27$
$O_2 + 4H^+ + 4e^- \rightarrow 2H_2O$	$+1.23$
$O_2 + H_2O + 2e^- \rightarrow HO_2 + OH^-$	-0.08
$PbSO_4 + 2e^- \rightarrow Pb + SO_4^{2-}$	-0.36
$Sn^{2+} + 2e^- \rightarrow Sn$	-0.14
$Sn^{4+} + 2e^- \rightarrow Sn^{2+}$	$+0.15$
$Fe^{2+} + 2e^- \rightarrow Fe$	-0.44
$Pb^{2+} + 2e^- \rightarrow Pb$	-0.36
$I_2 + 2e^- \rightarrow 2I^-$	0.54

Index

dispersion interaction, 122
Dulong and Petit, 85

Einstein, Albert, 84, 85, 89, 93
electrode, 58
electromagnetic radiation, 82, 83, 86, 88
electronegativity, 112
energy
    as heat, 2, 11, 12, 14–23, 25–27, 29, 32, 34, 83–85, 94, 128, 129
    hydrogen, 99, 105, 109
    internal energy, 11, 15, 17, 39
    photon, 85, 88, 89, 103, 134
energy level, 30, 31, 82, 88, 91, 92, 97, 102, 109, 112, 115, 117, 136
enthalphy, 128–132
entropy, 21, 129
equation
    Arrhenius, 69, 73
    Schrödinger, 82, 90, 92–95, 97, 98, 102, 105
    van der Waals, 10, 128
    virial, 10, 128
equation of state, 2, 3, 10, 122, 128
    perfect gas, 3
    van der Waals, 122

First Law, 11, 15, 16, 20, 21, 30
formation
    bond, 105, 106
    Gibbs free energy, 129
    rate of, 61, 63, 67, 73, 75, 132
    standard enthalpy, 18, 19, 30, 36, 55, 56, 128, 129, 132, 137
freezing point, 49–51, 131

Gibbs free energy, 33, 34, 38, 39, 58, 129, 130
    and equilibrium constant, 35, 40, 41, 55, 56, 59, 73, 77, 129, 130, 132

half-reaction, 57
harmonic oscillator energy, 85, 91, 116

heat, 2, 11, 12, 14–23, 25–27, 29, 32, 34, 83–85, 94, 128, 129
heat capacity, 14, 15, 18, 19, 29, 83, 85, 128, 129
Heisenberg, Werner, 90, 91, 93
Henry's Law, 47, 48, 130, 131
Hund's Rule, 103, 109
hybrid orbital, 108
hydrogen bonding, 122, 123, 125
hydrogenic atom, 98–100, 102

induction forces, 121–123
interaction, 10, 120, 121, 124
    dipole–dipole, 121
    dipole–induced dipole, 121
    van der Waals, 10, 121–123, 128
    dispersion, 122
internal energy, 11, 15, 17, 39
isolated system, 21, 27, 32, 33

Joule–Thomson, 10

kinetic energy, 7, 11, 71, 85, 135

Lindemann mechanism, 77
Lyman Series, 89, 101

molality, 51
moment of inertia, 115, 117, 119, 136

Newton, Isaac, 83
nonpolar molecules, 121–123

orbital, 82, 88, 99, 100, 102–113, 136
    hybrid, 108, 113
    moelecular, 104, 108, 109, 113, 136
osmotic pressure, 49, 52, 53, 131

P-branch, 118
partial molar Gibbs energy, 39
pascal, 2, 4
Paschen Series, 89, 135
Pauli, Wolfgang, 102, 103
phase rule, 43, 44
photoelectric effect, 83–86